BEES AND THEIR KEEPERS

LOTTE MÖLLER

Bees and Their Keepers

A Journey Through Seasons and Centuries

Translated from the Swedish by Frank Perry

ABRAMS IMAGE | NEW YORK

FIRST PUBLISHED IN GREAT BRITAIN BY MACLEHOSE,
AN IMPRINT OF QUERCUS PUBLISHING LTD, LONDON, IN 2020

FIRST PUBLISHED AS *BIN OCH MÄNNISKOR*
BY NORSTEDTS, STOCKHOLM, IN 2019

LIBRARY OF CONGRESS CONTROL NUMBER: 2021930591

ISBN: 978-1-4197-5114-1
eISBN: 978-1-64700-139-1

DESIGNED BY ANNIKA LYTH, LYTH & CO
PRINTED AND BOUND IN CHINA

10 9 8 7 6 5 4 3 2 1

ABRAMS IMAGE BOOKS ARE AVAILABLE AT SPECIAL DISCOUNTS WHEN PURCHASED IN QUANTITY FOR PREMIUMS AND PROMOTIONS AS WELL AS FUNDRAISING OR EDUCATIONAL USE. SPECIAL EDITIONS CAN ALSO BE CREATED TO SPECIFICATION. FOR DETAILS, CONTACT SPECIALSALES@ABRAMSBOOKS.COM OR THE ADDRESS BELOW.

ABRAMS The Art of Books
195 Broadway, New York, NY 10007
abramsbooks.com

CONTENTS

PART 2

The keeping of bees is like
the direction of sunbeams

Henry David Thoreau (1817–1862)

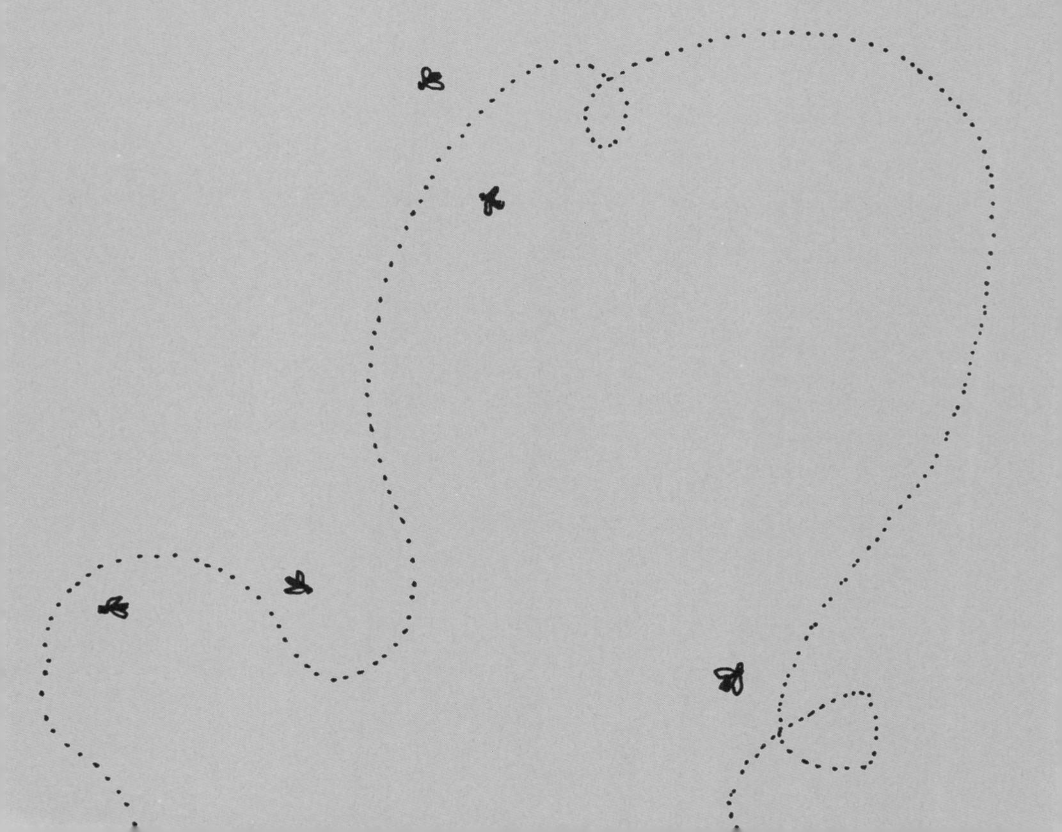

THE MIRACLE OF THE WAFER IN THE BEEHIVE

In most ancient cultures the bee was associated with religion and the divine. A symbol of virginity in medieval Catholicism, bees were believed to be parthenogenetic and many edifying legends associated them with the Holy Virgin. One of them concerned a farmer's wife who, instead of swallowing the wafer that the priest had given her during mass, had hidden it under her tongue before taking it home and stuffing it inside the log hive (that was how they made hives at the time) to attract more bees and therefore produce more honey and wax. But when she and her husband opened the log they discovered that the wafer had been miraculously transformed into Mary and the Baby Jesus.

SAMUEL LINNAEUS – THE BEE-KING FROM THE SWEDISH PROVINCE OF SMÅLAND

The honeybee got its Latin name *Apis mellifera*, which means the honey-bearing bee, from Carl von Linné, known outside Sweden as Linnaeus. But as Samuel, his younger brother by eleven years, would point out, bees did not carry home ready-made honey, rather the nectar that they would then process to make it. Its name should really be *Apis mellifica*, the honey-making bee. The rules of nomenclature did not, however, allow for this change to be made.

Samuel Linnaeus was born in 1718 in Stenbrohult. He took over from his father as vicar before becoming a dean, and eventually one of the pioneers of Swedish agriculture. By the time his book, *Kort men Tillförlitelig Bij-Skötsel* (A Brief but Reliable Guide to Beekeeping), was published in 1768, he had been keeping and studying bees for thirty years. It had a huge impact and is still worth reading.

In Egyptian mythology bees came from the tears of the sun god Ra when they fell on the desert sands. According to Greek and Roman authorities, including Virgil, they came into being in the rotting carcasses of oxen. This belief was called "begunia" and survived into the Middle Ages and even later. It was not until the eighteenth century that it was understood that the queen bee laid eggs after mating.

INTRODUCTION

The happiness of the bees and the dolphin is to exist.
For man it is to know that and to wonder at it.

Jacques Cousteau (1910–1997

The Roman writer Pliny believed the honeybee was the only insect created for the sake of man, a view that would prevail for a very long time and that can still be encountered today. Bees do not just give us honey and wax; they also set an example in terms of hard work, altruism and how to construct an efficient society. "These winged Aeolian harps can provide us with amusing company during our leisure which leads to wholesome observations and a refined mood," wrote the Swedish parson Fredrik Thorelius in the mid-nineteenth century, and his sentiments were typical of the period. Just as typical, though of a later era, is this observation by the Danish writer Jørgen Steen Nielsen made in 2017:

" We fancy ourselves the most intelligent of creatures. But intelligence encompasses many different things, including the ability to ensure a community's survival and its stability by employing a capacity to listen, collaborate and focus on the common good. If we fail to learn more of these qualities from the bees, who have far greater experience, we will lose first the bees and then ourselves. "

In the past keeping bees was an unquestioned part of subsistence farming. As was the custom at that time, Karl-Bertil and Anna Lovisa Johansson in Södra Vi protected their hives against rain and foul weather with hoods made of the bark of a spruce.

But bees exist no more for our sake than any other element of nature does. We, however, have made ourselves dependent on them. We can get by without their products if necessary, but most of what we eat derives from crops that are pollinated by insects, including the honeybee. And yet we have made things so difficult for them that their very survival is now in doubt. How did that happen?

In the past, agriculture was more diverse; the natural world was rich in different species and if you lived on the land, beekeeping was part of your existence. But as heather moors and heaths, pastureland and the places people lived have been replaced by plantations of conifers, while flowery meadows have been turned to arable land or simply cleared, rich sources of nectar and pollen have disappeared. The countryside has been depopulated, and fewer people keep bees.

The monocultures of today, oilseed rape cultivation in particular, provide huge amounts of nectar for a few weeks but nothing for the bees to live on over the rest of the summer. The chemical pesticides used in agriculture kill not only weeds, fungi and harmful insects but also honeybees, bumblebees, butterflies and other insects and the birds that live on them. The demand for profitability imposed by our society has made a mess of things – for ourselves and for many other creatures, including bees.

But there are opposing forces. As a result of books like Maja Lunde's *The History of Bees*, an awareness of the vulnerability of both bees and humans has grown. In recent years the sorts of people who keep bees have changed enormously. Women, young people, immigrants, graduates are acquiring bees on a scale that would have been inconceivable thirty or forty years ago when beekeeping was anything but fashionable. Beginners' courses are often oversubscribed. An increasingly popular urban form of beekeeping does not solve the pollination problem in agriculture and fruit-growing, but it does lead to curiosity and a greater awareness. There are now alternatives to conventional beekeeping practice that do not prioritise honey production but rather the survival and well-being of the bees. Keeping bees has become both essential and exciting in an entirely new way.

But it was not for the honey, the pollination or the survival of bees that I became a beekeeper in the 1980s. Bees just happened to me. It all began when as a freelance contributor to radio programmes I was asked to put together a feature for the Swedish midsummer holiday. Why not one about beekeepers? The year before, my friend Annicka Lundquist got a colony for her little summer cottage in Småland and became the first beekeeper I knew personally. In those days beekeeping was regarded by most as a quaint hobby practised by elderly men in the countryside, and in smaller towns and villages – retired teachers, local shop managers, farmers and stationmasters. Female beekeepers seemed to be as rare as female fire fighters. My idea for the radio programme was to interview Annicka as something of a pioneer, as well as a couple of beekeepers of the more traditional kind, the old boys of the beekeeping world, with some accompanying buzzing in the background.

It was then that I heard about John Larsson in Klagshamn. Every time the Malmö police were called out because a swarm had settled on a balcony or some other unsuitable site, the officer on duty would ring him and he would head over to deal with it. I got hold of his telephone number and he agreed to take me along the next time it happened.

It was in the car park outside Mobilia department store that we met for the first time. There he was in the bright sunlight, with sleeves rolled up and a small beret covering his bald patch, shooing a swarm of bees away from a parking meter and into a straw basket, assisted by his wife Inga, and recounting marvellous stories about the swarms they had captured in chimneys, from the masts of boats and around flagpoles. The couple were surrounded, albeit at a safe distance, by curious onlookers.

"Don't they sting?" asked someone newly arrived on the scene.

"No," John said, "bees and policemen are as nice as can be as long as you don't irritate them."

Bees were crawling all over his arms, his throat and his face and even into his ears. But he refused to pay them any attention. The important thing, he explained, was to get the queen into the basket. The others would placidly follow.

John Larsson was not just a wizard at capturing bee swarms. He was also a marvellous entertainer. Jana, an acquaintance of mine, found herself one hot June day at the women's section of the Ribersborg outdoor baths in Malmö when the air was suddenly filled with buzzing bees. Someone called the police and a short while later John Larsson appeared. "As mature ladies, we jumped straight in the water or threw ourselves onto our bellies when a man suddenly appeared, but once he set about capturing the bees and started telling us about them it wasn't long before we were crowding round him patting his forehead dry with cold towels."

"How do you know which one she is?" a plucky little girl said.

"By her crown," he said, but then Inga twisted his ear.

"You mustn't lie to children. You can recognise the queen because she has a long rear."

What a show they put on. The tape recorder kept whirring and it turned into a pretty good programme, thanks to the Larssons in particular.

The writer Lars Norén also took part because he had written a novel called *Biskötarna* (The Beekeepers). It was about life among the down-and-outs in Stockholm and had nothing to do with bees, but even so I managed somehow to make it work. And I stayed in touch with Inga and John once the programme had been broadcast. That was not the case with Lars Norén, just in case you were wondering.

One July morning the following year, John rang me to let me know I could come and pick up the hive he had built for me.

"And there's bees in there as well."

Help! It had been the beekeepers, not the bees, that had interested me when I was making the programme. Their technical terms – honey super, brood room, queen excluder, cast swarm, landing board – sounded both lovely and magical and they had this marvellous way of talking about their buzzing friends. I told him I could not possibly accept his gift, and besides, I was actually afraid of bees. But John just laughed and said I was bound to get the bee bug as soon as the hive had been set up in my garden. Inga and he would show me how to care for my bees, and if I had any problems all I had to do was give them a call.

I had no choice. I drove off to Klagshamn to fetch the hive and its tens of thousands of inhabitants, and later that summer I would be given another, and that had bees inside as well. A new chapter in my life had begun, and even if it would be anxiety-ridden from time to time, I will always be grateful to John. There was so much for me to learn. How to handle a smoker, how to distinguish between queen, drone and worker-bee cells, how to watch for signs that a colony is about to swarm, how to uncap honeycomb, how to extract honey, how to install partitions in hives and prepare them for winter.

I had also to learn about the relationship between flowers and bees and I began to look at my garden and its surroundings in a new way: with an eye to pollen and nectar sources. Maple: good. Lime and robinia: very good. Raspberry and currant bushes: good. Willow and hazel: good. Sallow, thyme and lavender: the best. The perfect lawn: completely useless, but if it has a lot of clover and other so-called weeds, that is quite a different matter. John was right. My biosphere had been expanded and living in close proximity to my bees became an exciting if nerve-racking experience.

"You should let them crawl over you so they learn your smell; then they'll only sting if they get stuck," John said.

Not that I dared, of course. That first year my protective clothing consisted of an old dressing gown, a bee hat with a veil and rubber gloves. I would subsequently acquire a white overall which looked more professional.

But even getting to know other bee people lent my life an added dimension. You know the way that works. If you're suffering from a stiff neck or a heel spur, you're suddenly surrounded by other people with the same affliction. If you've just had grandchildren everyone you come across has just

My little apiary with the hives that John Larsson built.

become a grandparent. And all of a sudden there were bee people everywhere, or the children, spouses or friends of beekeepers at least. Conversations with other people, with people you knew as well as with perfect strangers, would sooner or later involve bees and honey.

I also discovered that just as bees organise themselves into colonies, beekeepers set up associations. I became a member of Södra Sveriges Biodlares Förening (S.S.B.F., the Beekeepers Association of Southern Sweden) simply because the Larssons belonged to it. The association was a breakaway group from the mighty Sveriges Biodlares Riksförbund (S.B.R., the National Beekeepers Association of Sweden) and had been set up in protest against the national association's insistence that all the honey produced by members had to be sent to the tapping facility in Mantorp in the province of Östergötland, to be mixed together into a homogenised Standard Swedish Honey. Dark and intense heather honey, firm white rape honey, pepperminty linden honey, creamy raspberry honey and thyme honey from Öland were all to lose their defining qualities as they were transformed into a standardised product. This was about as brilliant an idea as blending all the wines of France from Bordeaux to Burgundy, from the Loire to the Rhône, into a homogeneous vin Français.

S.B.R.'s argument was that customers got confused when honey looked and tasted different. On the contrary, the S.S.B.F. insisted, one of the great joys of honey was that it tasted and looked different depending on where the bees had gathered their nectar. In addition, all honey sales were supposed to be made through the national association which the S.S.B.F. regarded as a grave violation of individual liberty. It was the bureaucrats versus the individualists!*

When I joined, the S.B.R. had, however, begun to reverse its policy on whether varietal honey should exist and the S.S.B.F.

§ 12.
Ordföranden betonade att föreningen skall vara klar och redo att träda i verksamhet i god tid till nästa honungssäsong, om inte dessförinnan Sveriges Biodlares Riksförbund genom ändring av sin nu intagna ståndpunkt, erkänner sina medlemmars rätt att själva sälja sin producerade honung och själva uppbära likviden därför.

From the minutes of the Beekeepers Association of Southern Sweden's first meeting held in November 1955. They include a demand that the National Association should allow its members to sell their own honey and keep the proceeds.

had already seen its best days. A few of the older members had left the latter association after the fields around their apiaries had been sprayed and their bees died. There were very few new members and at each annual general meeting the minute's silence for members who had died seemed longer and longer. In the end even the association itself died a death, because not enough members turned up for meetings to allow any decisions to be made.

John Larsson was among those who passed away. It was very sad, and it seemed as if it was not only his family and

*Nowadays the national association – Biodlarna (the Swedish Beekeepers Association) as it is called today – has a very different policy on varietal honey. They seek both to protect the different kinds of honey and to publicise their existence, and even encourage celebrity chefs to prepare dishes with honey as one of the ingredients. The facility at Mantorp was sold long ago.

many friends who mourned him but my bees as well. There used to be an old custom that bees had to be informed if their keeper died, else they became too difficult to handle or simply died themselves. I should have told my bees that John had died. They had grown accustomed to his visits and became increasingly touchy and easily irritated. If I mowed the grass they went crazy and harvesting their honey became an ordeal.

After keeping bees for ten years, I gave them away. I had begun commuting on a weekly basis between Stockholm and Lund, and that was not compatible with my responsibilities towards them. In some respects it was the right moment. The loathsome Varroa mite, *Varroa destructor*, which makes the bee's wings deformed and underdeveloped and stunts their lower bodies, and is capable of killing an entire colony of bees, had not yet arrived in Sweden, but it was making its way here at top speed. That was one scourge I could avoid having to worry about. Being without bees was nice. For a while at least.

Soon enough, though, the garden began to feel empty, and I was thrilled when Rolf and Margrethe Jönsson from the flower shop on the other side of the road asked if they could place two of their hives in my garden – they had no room left on their allotment. Imagine being able to follow the comings and goings of bees without being responsible for them. But after a few years both their colonies died and that was the end of that pleasure. Once again I was missing the buzz around the hives, those shining sacs full of pollen the bees would come home with, the joy when they appeared around the hive entrance on the first day of spring. Maybe I ought to start keeping bees of my own again? There was a new kind of hive, the top-bar hive, that would suit me better now that I no longer have the strength to lift as much. But come to think of it that just wouldn't work. There's been a lot more traffic passing outside my house in the last few years and the ground vibrates when heavy vehicles roll by. Bees don't like that kind of thing. That might be why the Jönssons' bees died.

Instead I began to read about the social and cultural history of beekeeping and that turned out to be at least as exciting as having bees, though in a different way, of course. And there was so much of it! Even the Ancient Greeks and Romans wrote about bees, as did Saint Bridget, Olaus Magnus and Linnaeus' younger brother Samuel, along with Voltaire, Réaumur, Shakespeare and Selma Lagerlöf. And many of the old texts, the manuals in particular, were written with so much personal feeling, such a sense of wonder and in so painterly a language that they are poetry, pure and simple.

"The Queen Bee is a Creature both Lovely and Magnificent"

A phrase such as that can brighten up a whole day. It can be found in *Nödigt Tractat om Bij* (An Essential Dissertation on Bees) by Mårten Triewald (1691–1747), one of the founders of the Royal Swedish Academy of Sciences, the man who introduced the steam engine to Sweden – after studying its construction in England – and an authority on bees.

But even though I had read only a tiny fraction of 1 per cent of 1 per cent of everything that has been written about bees, I soon felt like a worker bee whose nectar sac is full. While she will fly home to the hive so that the result of her efforts can be transformed into honey, I sat down at the computer to write this book.

PART ONE

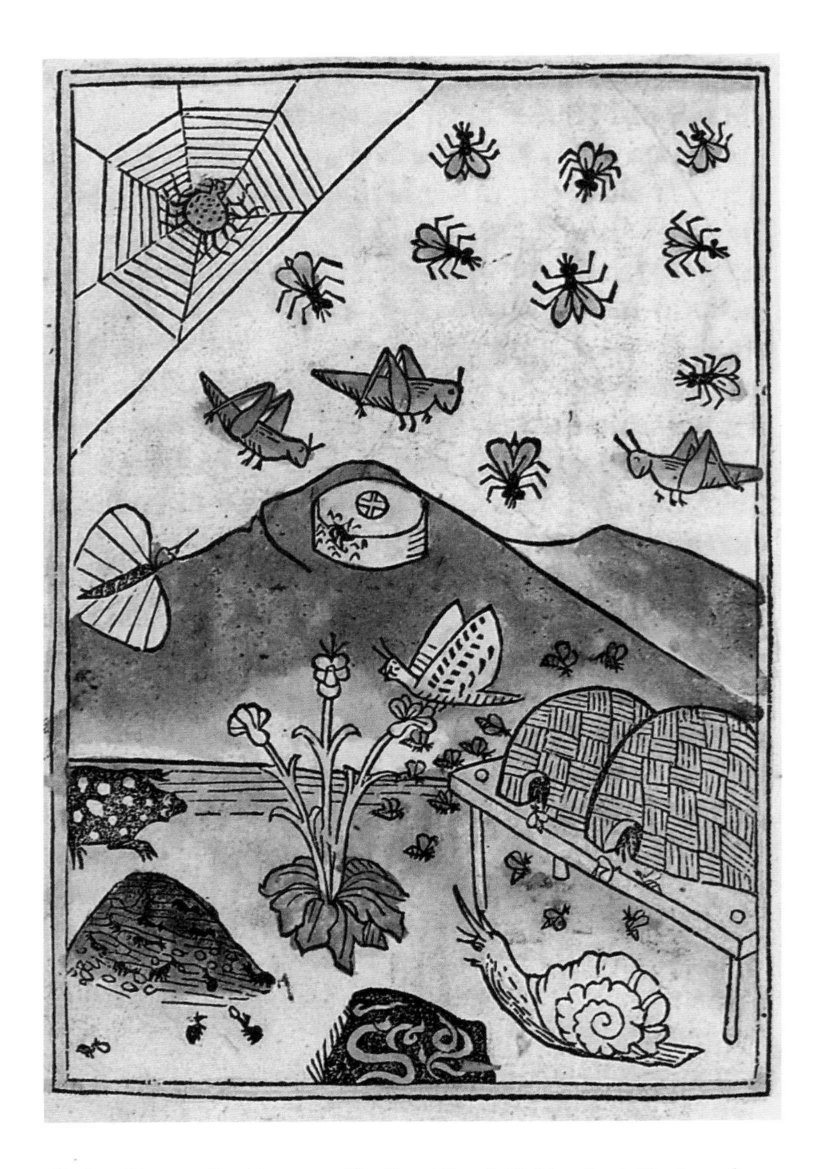

Medieval bee enemies. Some are still a threat though that hardly applies to toads – if only because they have become so rare. From Konrad von Megenberg's Buch der Natur *of 1481.*

JANUARY

A Winter Memory

and

A Description of the Enemies of Bees, Old and New

It is snowing and I am wondering how my bees are doing out there in their hives. Have they got any of their winter food left? Can they keep the heat going? It's all a bit worrying. Every year it's the same. Winter is either too long or too short. Too mild or too cold. When you listen to experienced beekeepers it can sound as though something is bound to go wrong in wintertime. There don't seem to be many good springs or summers either. They are either too hot and dry or too wet and chilly.

This has to be one of the bonuses of beekeeping as a hobby, though: you've always got something to be concerned about, not too much and not too little. It helps to keep your real problems at bay. As it says in *Stora biboken* (The Big Bee Book): "Whatever the worries, troubles or annoyances you are facing, you forget them the moment the apiary and its maintenance require your attention." Birgitta Stenberg says something along the same lines: "The bee colonies would help keep me going when all the injustices of the world were pouring in via television or the radio. I would shut out these torments by leaving the house for the realm of the bees in their hives."

BIRGITTA STENBERG (1932–2014)
is best known for her daring autobiographical works, but she also kept bees on the Swedish island of Åstol with her husband Håkan, and wrote a highly personal and yet very instructive book, *Allt möjligt om bin* (Anything and Everything about Bees), which she illustrated herself.

All the same, it is still too early in the year to open the hives. That would be catastrophic for the bees, which are now clustered in a rotating ball to keep in the heat. Disturbing bees when they're in this state can lead to them developing a form of dysentery. So the key is to tackle your anxiety some other way.

You could, for example, feed the great tits. These otherwise delightful birds can cause a great deal of trouble when hungry. If they come across a beehive, they will sit on the landing board and peck at it until some of the still-dazed guard bees appear to find out what is going on. They are pounced on and eaten then and there. And then peck, peck once again and yum yum. If this goes on for any length of time the entire cluster can break apart and the colony may die due to dysentery and the loss of heat caused by the great tits' harassment.

So you have to keep the birds well fed with seeds and tallow balls – and even that may not be enough. They might decide bees are a delicacy they simply cannot resist, no matter how much you have fed them.

Nature is wondrous strange after all, and completely unsentimental.

How do you trick a great tit?

Great tits are not one of the major problems confronting today's beekeepers; perhaps they were once more numerous? In beekeeping manuals of the past there are many accounts of their thoughtless rampaging, along with advice about how to stop them. But only one writer – the extraordinarily knowledgeable eighteenth-century Dane Esaias Fleischer – mentions feeding them specifically. A box containing a bit of meat or tallow placed beside the hives will readily entice "these greedy and audacious birds", he writes. But then the problem is what to do with them. Shooting close to a hive is not a good idea, Fleischer points out.

The great tit, pretty but untrustworthy. To a bee at least.

Shooting great tits! But Fleischer had a more humane proposal: If you attach a piece of cloth, in scarlet or some other bright-red colour, above the landing board, no great tit will dare alight on it. Here is another piece of Danish advice:

> "Great tits behave atrociously towards bees. They sit in front of the entrance to the hives and use their beaks to peck at it. A few bees will rush immediately to the spot and the birds can snatch one bee after the other. Then they fly to the next hive and devour the bees there. These small birds have developed this bad habit by eating dead bees that have been evicted by the living ones. For this reason it is a good idea to keep the areas around the hives neat and tidy."

From *En nyttig bog om bier* (A Useful Book about Bees)
by Hans Herwigk, 1649

Great tits were also a problem in the Swedish province of Småland:

“Birds often cause bees a lot of trouble, particularly the ones known as Great Tits. In autumn and winter, and even in spring, they become most brazen and peck away at the entrance or the edge of the straw above it, and then as soon as a bee or two has emerged, they are snapped up and devoured; they disturb the bees in winter by pounding on the skep and luring them away from the cluster they have formed so they soon freeze and die.”

From *Kort men tillförlitelig Bij-Skötsel* (A Brief but Reliable Guide to Beekeeping) by Samuel Linnaeus, 1768

Writers of fiction have also described the mischief the great tits get up to in winter. There is a wonderful passage in *Gösta Berling's Saga* by Selma Lagerlöf. She clearly knew a lot about bees:

“Near the opening of the beehive a great titmouse was engaged upon a perfectly fiendish trick. He must have his dinner, of course, and he tapped, therefore, at the opening with his sharp little beak. Inside the hive the bees hung in a big, dark cluster. Everything within was in the strictest order. The workers dealt out the rations, and the cup-bearers ran from mouth to mouth with the nectar and ambrosia. With a constant creeping movement those hanging in the middle of the swarm changed places with those on the outside, so that warmth and comfort might be equally divided.

They hear the titmouse tapping, and the whole hive becomes a buzz of curiosity. Is it a friend or an enemy? Is there danger to the community? The queen has a bad conscience, she cannot wait in peace and quietness. Can it be the ghosts of murdered drones that are tapping out there? “Go and see what it is,” she orders Sister Doorkeeper, and she goes. With a “Long live the Queen!” she rushes out and ha! the titmouse has got her! With outstretched neck and wings, trembling with eagerness, he catches, kills, and eats her, and no one carries the tale of her fate to her companions. But the titmouse continues to tap and the queen to send forth her doorkeepers, and they all disappear. No-one returns to tell her who is tapping. Ugh! it is awful to be alone in the dark hive – the spirit of revenge is there. Oh, to be without ears! If one only felt no curiosity, if one could only wait in patience! ”

Bengt Hallgren (1922–2017), a rather more recent writer from Värmland in Sweden, also wrote about greedy great tits.

“ The great tits perched on the landing board and pecked away until the curious bees appeared. Then the great tits snapped them up. Though they never ate the lower body where the sting is. Grandfather experimented with landing board covers to stop them knocking. He also placed a brick in front of the entrance but that meant the bees were at risk of suffocation. Pålsson cobbled together a tit trap. He captured a score of great tits using a bit of tallow. But Grandfather could not bring himself to kill the culprits, he let them loose instead. ”

From *Farfars honung* (Grandfather's Honey), 1969

Great tits are far from the only creatures beekeepers are warned about in the literature of the past. They were also to look out for swallows, swifts, woodpeckers, storks, ducks, turkeys, peacocks, wasps, bumblebees, grasshoppers, ants, dragonflies, wax moths, other moths including the death's-head moth, spiders and robber bees. Frogs and toads could be concealed in any long grass growing beneath the hives and they would poison the bees with their breath, or else snatch them out of the air.

Among mammals, bears and mice were notorious pests, though badgers and foxes could also cause problems by overturning the hives.

"When bees are attacked by wasps they utter a piteous sound as though they were signalling their distress to their owner."

Samuel Linnaeus

It is in the eighteenth century, during the Enlightenment, that another enemy of bees is first mentioned, and one even worse than all the others: human beings. In *Nya svenska economiska dictionnairen* (The New Swedish Economic Dictionary) of 1779 this is explained inasmuch as many beekeepers "both through careless handling [. . .] and by ruthless killing and slaughter display their ingratitude towards these most beneficial of insects". What "ruthless killing and slaughter" could mean is described in *Lantmannens uppslagsbok* (The Countryman's Almanac) of 1923:

"Bee slaughter is the barbaric treatment meted out to bees by our ancestors and, in more recent times, by some straw-hive beekeepers. With the arrival of autumn some hives would be condemned to slaughter which involved digging a hole in the ground, sulphur was then put into the hole and set alight, whereupon the hive was placed above the hole. When the colony was deemed to be dead, the hive was removed and the contents harvested, a mixture of the brood, pollen, and the bees along with the excreta they had passed on the combs and in the honey in their death throes. This is unlikely to have been made more appetising when the hive had also been sealed both internally and on the outside with cow dung."

A rather less sentimental description, the childhood memory of an elderly man, was recorded in Uppland, Sweden, in 1930:

"When they took the honey from the bees, a hole was dug and lined with linen cloths dipped in sulphur which were placed on sticks and then in the hole. They then set fire to it and placed the hive above it and all the bees were killed. Then the combs were wrapped

Honey being harvested using sulphur; oil painting by C. G. Bernhardsson (1915–88). He lived on the island of Skaftö and in his many paintings portrayed everyday life in the Swedish province of Bohuslän at the turn of the previous century.

in thin linen cloths and these were tied to sticks so the honey could flow out into containers placed beneath. Water was then thrown on the combs and allowed to drain away. The wax was melted and filtered. A wonderful honey drink was brewed with hops. The wax was used by tailors to wax their threads and to make wax sticks (candles) for the Christmas tree. ”

Until the advent of the stackable frame hive at the beginning of the twentieth century, this was the usual way of obtaining honey throughout Europe – despite persistent scientific condemnation, despite new types of hives that allowed the honey to be removed without destroying the bees – one of which was designed by Samuel Linnaeus – and despite the work of associations such as the 1830s U.S. bee-rights movement Never Kill a Bee and Thomas Nutt's popular 1845 book *Humanity to the Honeybees.*

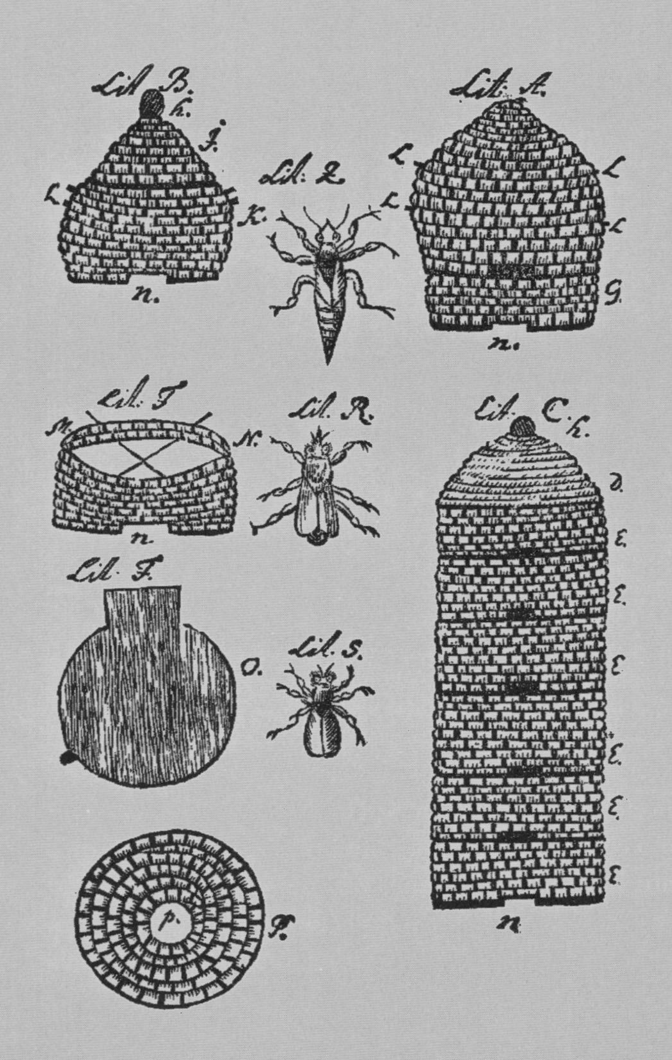

SAMUEL LINNAEUS' HIVE

He referred to it as a "colonie". The final construction can be seen in the image bottom right. It was woven from straw and hazel and sealed with a paste made of cow dung, clay and water. New floors – rings – could be added from below once the existing ones were full. This makes it a predecessor to the Warré hive designed in the 1950s (p.197f.) This form of hive meant the honey could be obtained without killing the bees as the upper rings which contained only honey could simply be lifted off.

Many people blamed the ignorant commoner as the reason such cruelty persisted. But even many knowledgeable beekeepers preferred destruction to harvesting. The chief beekeeper at Torup Castle in Southern Sweden, who was a celebrated authority in the eighteenth century, had tried both techniques and thought that removing the honey was "foolishness". Destroying the colony meant healthier bees, fewer pests and far more honey, in his view, though it was of course important to not mix together honey, wax and dead bees but to take pains to keep them separate.

Even academic experts supported the killing of bees. Lars Laurel (1705–93), professor of theoretical philosophy at the university of Lund and member of the Royal Swedish Academy of Sciences, maintained that bees reproduced so quickly that there could never be enough food for them. This made it essential to cull them, in the same way that if too many calves were born on a farm they would be slaughtered.

> "But some people are so soft-hearted they cannot bring themselves to kill the bees. It pains them as much as if they were rational creatures [. . .] But just as the flower knows nothing about its growth, so too the bee is ignorant of its actions and therefore of its death and the killing of bees should touch one no more than the cutting of the grass on the ground which becomes necessary in autumn."

What it felt like for what Laurel called a soft-hearted person to kill their bees is made clear by the English farmer's wife Ann Hughes in her diary entry from the end of the eighteenth century:

> "It do grieve me to kill the poor things, being such a waste of good bees, to lie in a great heap at the bottom of the hole when the skep be took off it; but we do want the honey, using a great lot in the house for divers things."

Human beings are still the greatest enemy that bees face, and far more dangerous now than when we were destroying them with sulphur. Entire apiaries were not, after all, being destroyed at that time, but ever since we began spraying crops with pesticides the prospect of extinction has come to haunt the beekeeping world.

"Vast numbers of bees have died on the island of Ekerö leading to the annihilation of some 300 colonies to this point," the Swedish newspaper *Aftonbladet* reported in June 1938. "The cause was discovered to be the ill-advised spraying of strawberry crops by gardeners using a new product against raspberry weevil, and flea beetles that contains powdered arsenic." James Whorton's 1974 book *Before Silent Spring* has much to report on the use of arsenic on bees, a major problem in the U.S.A. in the nineteenth century.

The use of arsenic on flowering plants that are visited by honeybees and bumblebees was prohibited (in Sweden) in 1945. The very year that D.D.T. was first marketed. It was launched as a miracle product for agricultural and domestic use but turned out to have a dreadful effect on the environment as a whole, and is now prohibited in most countries. It was replaced by the neonicotinoids, which are amongst the main causes of the current mass deaths of bees and other pollinators. One of their effects is that the bee, even if it does not die on exposure, becomes disoriented and cannot find its way back to the hive, or is so weakened that it becomes vulnerable to diseases and parasites. In 2018 the E.U. finally prohibited three of the neonicotinoids. The weedkiller glyphosate, found in products such as Roundup, has also been identified as a threat. It also leads to bees losing their ability to navigate. The affected bees then take the poison back into the hive, and so there are large amounts of it present in much of the honey that is sold.

The picture does seem a little rosier in relation to other threats faced by bees. Some animals are nowadays so rare that they have been made protected species, while the danger posed by others has been exaggerated. It is no myth, though, that woodpeckers peck their way into the hive to feast on the bees' honey, and if the hives are made of expanded polystyrene or other kinds of plastic rather than wood, that makes it even easier. But this must be offset by the fact that the number of woodpeckers has fallen drastically.

Bears, on the other hand, have grown in number since the nineteenth century when they were all but exterminated. There are currently as many as 3,000 bears in Sweden. Good news for people who care about animals in the wild, but not for beekeepers (nor for sheep farmers and reindeer herders) in those parts of Sweden – the very centre and the northern regions – where the majority of bears are to be found. They are known to love the taste of honey and can devastate entire apiaries. Bears are a problem for beekeepers in many other countries as well, including Finland, Norway, Italy, Austria, Ukraine, France and the U.S.A. Electric fences provide the best protection, but if the bears are really hungry for honey they will dig their way beneath them.

“On Monday the county council confirmed that a bear had smashed some beehives in Hagby. The apiary affected is run by Göran Eriksson who, despite his twenty-five-year career as a beekeeper, could not at first believe that the damage had been done by a bear. 'I had no idea there were any bears in the area,' he says. 'That's why I just couldn't believe my eyes.' Four of the sixteen hives in the apiary were struck and two of the colonies could not be saved.”

Upsala Nya Tidning, May 13, 2015

“The crime spree started when staff at the Pennsylvania university found that one of the bee hives at the Center for Earth and

Environmental Research had been ransacked and 'heavily licked' [. . .] University staff was able to reassemble the damaged hive, but the crook came back the very next night to once again dip his undoubtedly very sticky paws back in the hive for more honey [. . .] While they have no solid proof or photographic evidence of the criminal mastermind at work, the university is confident that the thief is a 'naughty bear'. ”

Time, March 29, 2019

Unlike bears, mice can be found pretty much everywhere. They love to steal into the cozy warmth inside beehives in the autumn, when food is constantly being served. Sometimes the bees manage to sting them to death and embalm them in bactericidal propolis. Barrier grating, netting and reducing the height of the entrance are all recommended as protection against mice.

Ants are another major cause for concern. "Once they get inside the hive they can attack the bees," Samuel Linnaeus wrote. To stop them getting inside "as though they were on a main road" the legs of the bee bench (contemporary hives were placed on benches) could be placed in bowls of water. This is still done today with the feet of modern hives, although they can also be encased in solid plastic. Some beekeepers also sprinkle cinnamon or scatter crushed eggshells.

A curious technique that Esaias Fleischer found to be best after experimenting with a great number of different methods was one he read about in a contribution by a "Mr Boetius in Wester-Aas" in *Abhandlung der königlichen Schwedischen Academie der Wissenschaften* (A Dissertation of the Royal Swedish Academy of Sciences). He recommended tying used and therefore still odorous fishing nets around the hives, which meant the ants were unable to come anywhere near them.

Mr Boetius was in fact Jacob Boethius (1647–1718), a vicar in the parish of Mora in the see of Västerås. He was also a

Bears love the taste of honey. From the very beginnings of beekeeping right up to the present day that sweet tooth of theirs has been a constant cause for alarm in areas with resident bear populations. Aberdeen Bestiary c. 1200.

beekeeper, for a time at least, but the article about ants and fishing nets was all he wrote about bees. He became famous for his opposition to the Swedish Church Law of 1686, which made the king, an absolute monarch, the supreme authority of the Church of Sweden. He criticised the declaration of the majority of the young Charles XII and was condemned to death as a result. The sentence was commuted to life imprisonment and eventually he was pardoned. He refused the pardon and was transferred to the hospital at Danviken – yet another of the fascinating characters so frequently encountered in the history of beekeeping.

Wax moths prefer to lay their eggs in the wax cells of the beehive. The larvae gnaw channels through the wax and into the wood of the hives and frames and can cause enormous damage. The ancient Romans Pliny the Elder and Varro complained about them in antiquity. Esaias Fleischer recounts a rumour among the common people suggesting that the huge increase in the numbers of wax moths was the result of the Church no longer using beeswax candles, which had previously provided them with a form of fast food. The real cause, however, according to Fleischer, was the freezing winter of 1740, which had killed off a large number of bee colonies. This meant the moths were able to rampage unhindered through the hives.

And yet this still pestilential insect may one day be praised as an environmental hero. Only a year or two ago a Spanish scientist discovered that the wax moth also eats plastic and can break down polythene. Beekeepers with plastic hives may view the wax moth becoming a plastic moth as less desirable, however.

Robber bees – yet another eternal scourge of honeybees – are not a special kind of bee but our ordinary *Apis mellifera* once they have discovered that stealing honey from weak colonies is more convenient than producing their own. Samuel Linnaeus' advice was to give the bees under attack some sweet wine, Spanish or Portuguese were his suggestions, mixed with honey. This made them more courageous and better equipped to defend themselves against invasion. "One farmer used Strong Liquor instead of Wine and told me that its effect

Ants that have found honey. Is that the ant heap on the right and the honeycombs on the left, or is it the other way around? Northumberland Bestiary, c. 1250–60.

was beneficial." The Dane Hans Herwigk also believed that alcohol improved the fighting ability of bees. "Some of my neighbours take good old mead and melt a bit of honey into it and give it to their bees before sunrise on the Feast of the Annunciation, after which they remain invulnerable to robber bees for the rest of the year."

HANS HERWIGK

lived in Roskilde in Denmark and was the manager of the public baths, an unusual occupation for the author of a book on bees. Much later they would almost all be priests. Herwigk's *En nyttig bok om bier* (A Useful Book about Bees) of 1649 is the first Scandinavian manual for beekeepers and is packed with interesting and amusing observations.

If this failed you could take fresh honey directly from the combs and mix it with good wine and spring water, and on a morning pour the liquid over the wax combs of the colony under attack. The hive would be kept closed for the rest of the day, and then the bees were ready for battle once more. Currently the best advice is to reduce the size of the opening above the landing board, making it easier for the watch bees of the colony to drive off the robbers. But if any curious beekeeper cares to experiment with giving his bees a wee dram I would love to know what happens. Does it make them pluckier?

The list of potential dangers for bees does not end here. There are diseases and parasites such as nosema, European foulbrood, American foulbrood and tiny hive beetles. The worst of the lot is the Varroa mite. But on this point, dear readers, I refuse to go into more detail. A reader wishing to deepen their knowledge of the pest in question can consult any of the excellent handbooks that are currently available.

The time for a cleansing flight is still some way off in this depiction of life in the country in the month of February from Les très riches heures du duc de Berry, *an illuminated book of hours from the early thirteenth century. The bees are clustered together inside the hives and hopefully have a little honey left to live on. But the moment spring is in the air they fly outside to relieve themselves.*

FEBRUARY

A Sign of Spring Remembered

and

Older Descriptions of the Cleansing Flight and the Reactions of Neighbours

THOSE CUNNING DEVILS the great tits are sharpening their claws. The bees that have survived all the ills of winter have been out on a cleansing flight, which is another sign of spring. Having kept it in ever since the autumn they fly outside in a single body and relieve themselves the first day the air feels warm.

Until now I have missed this magical confirmation that my bees have survived the winter. I may have seen the traces left behind but that is not the same thing. And now it has actually happened! I was in the garden cutting back branches when I suddenly caught sight of the bees pouring out of the hives and then circling in the air. Absolutely incredible.

Bees on a cleansing flight are said to prefer something white as their goal – white as snow. This is why it is not a good idea to hang out the washing on the first day of spring because the spots of bee poo are almost impossible to get rid of. Not that I did, but I had been considering it. Air-dried washing smells divine, particularly the first wash of the year, only the washing machine was broken. A blessing in disguise. Shiny paintwork on cars is also supposed to attract them. But I don't have a car and therefore no paintwork to worry about. All I could do was enjoy the experience. The bees seemed in fact to be aiming at abandoned pieces of garden furniture that had been left out over the winter. Just as well, because now they will finally have to be painted.

Eventually they returned home, but not to rest. The queens have already begun to lay eggs and that means the workers really will have to justify their name. The larvae have to be fed, wax produced, cells built and fairly soon they will have to bring in pollen from hazels, willows and crocus flowers. It is all starting up again. The cleansing flight is what you might call the bees' New Year's Eve party.

The Joys of Spring or a Public Nuisance?

These days, authors on bees have little to say about the cleansing flight, if they refer to it at all. But there are many less recent accounts and they are often lyrical.

> "When the sun starts warming the hives at 9, 10 or 11 in the morning, the bees rush out in vast numbers, giving voice to a joyful tune about the end of winter and the approach of sweet spring, turning circles in the air to signal where they belong, releasing their winter plugs in flight or landing on the ground or elsewhere to cleanse themselves of the excreta accumulated during the winter and assuring their owners of the diligence they will show throughout the summer."

From *Hufvud-Grunderne i Biskötseln för Enfaldige Landtmän*
(The Principles of Bee-Keeping for Simple Countrymen)
by G. Natt och Dag, 1768

"Bees giving voice to a joyful tune about the end of winter and the approach of sweet spring, turning circles in the air to signal where they belong, releasing their winter plugs in flight."
Image from the fifteenth-century Tacuinum sanitatis casanatensis

Isn't that a marvellous description? Hard facts blended with a celebration of the joys of spring.

I am very taken with an account written in 1904 by the Irish priest the Rev. J. G. Digge for an entirely different reason. The pains this prudish Victorian takes to avoid mentioning what the bees are actually doing during their spring excursion are fascinating.

" A gladsome hour this is for the bee-man also; an infectious happiness. He knows now that the snow and storms, and all the frost and cruel winter hardships have failed to work their devastation within the little home which his foresight and loving care secured and sheltered before the falling leaves had left the branches bare. With each succeeding sun the bees in larger numbers move abroad – creatures 'fanatically cleanly', who will suffer much and long and yet refuse to sully the purity that their incessant care preserves within the hives. "

The joy a beekeeper may experience in relation to a cleansing flight is not, however, always shared by the neighbourhood. Many a conflict with a neighbour concerns bee-poo, no doubt also exacerbating an already tense relationship between the parties.

" This is the first time that environmental inspector Susanne Johansson of the West Blekinge Environmental Association has received a complaint about dirty bees. The people behind the complaint are neighbours of the beekeeper concerned. She has searched the law books but been unable to find any legislation that would cover a complaint of this kind. 'The bees were pooing on a car. That damaged the paintwork. They then pooed on the washing and a caravan,' says Susanne Johansson. "

Blekinge Läns Tidning, May 15, 2007

Dirty bees! On the contrary, bees are unusually clean animals. Still, touchy neighbours should be warned about the cleansing flight.

"Life at Kullabi's queen-breeding farm in southern Jönstorp is not entirely friction free. The neighbours have been complaining for the past few years and now a number of new complaints have reached the local authority's environmental committee. The complaints are not about the bees themselves but about their

Home to the hive after the cleansing flight. Time to get to work! Illumination from the thirteenth century

excrement. One neighbour got bee-poo on the front of his newly painted house. But the poo has also been found on cars, on roofs and on washing hung out to dry.

'I really didn't want to get involved in the row because bees are a tricky subject to complain about. But the final straw was when we got bee-poo in our food,' says one neighbour who wants to remain anonymous."

Helsingborgs Dagblad, September 8, 2015

An American apiary or bee-farm at the end of the nineteenth century

MARCH

A Feisty Californian Beekeeper Remembered
and
An Update on the Progress of the Killer Bee

IT IS MARCH 1984. The sun is shining over the San Joaquin Valley in California. I am here to visit friends in Hughson but when I am told there is a beekeeper in the neighbourhood, the priority has to be to find out if that is true. He does indeed exist and even likes to receive visitors at his Honey Farm.

As I arrive he is moving mountains of hive boxes that need repairing. It is obvious that this is no amateur beekeeper.

"Jack Beekman – B for bee, k for keeper and man because I am a man!"

That is how he introduces himself, before adding that today he is also a very happy man. Laura Belle, his wife, has just come home after a long stay in hospital following a complex leg fracture.

"She told me early this morning she was afraid she'd never be able to walk again," he says, "but then I said 'What hooey. Your ancestors were true Yankees who cleared the land, fought against Indians and made the long journey west, and with blood like that in your veins you don't just give up on account of an itty-bitty fracture'."

I go indoors to the living room to say hello. I find her lying on the sofa, a small, frail woman wrapped in a blanket. The family photos are lined up on the table beside the sofa. An only child, she grew up on this farm, she tells me, and it could sometimes feel rather lonely. Her parents grew almonds and oranges. The bees arrived with Jack, with whom she has four children. At least they've never had to feel lonely, she says.

She actually doesn't care that much for bees herself. But she is glad the business is doing well and that two of their sons have joined the firm. The farm is currently made up of 4,000 colonies, which is an astronomic figure by Swedish standards. And yet the Beekman's place is not considered one of the major apiaries in the valley. The really large-scale beekeepers have 20,000 colonies or more.

Like many other Californian beekeepers, the Beekmans practise migratory beekeeping. The hives are moved to where the plants are blooming most abundantly and where their services as pollinators are most required. Many of the hives are currently right here on the farm, however, because parts of the orange orchards owned by Laura Belle's parents still exist. There are so many scents in the air, so much buzzing! Other Beekman bees are foraging among the almond trees further along the valley. Afterwards they will be moved to the mountain slopes where the most extensive orange groves are planted and where the wild sage is in flower. Later in the season it will be the turn of the alfalfa. We grow it on windowsills for the sprouts; here it is a grazing crop.

Despite the favourable climate that has made California the leading bee state in the U.S., the beekeepers here have their fair share of worries, as it turns out. Among the things upsetting Jack Beekman is that the government is permitting

Jack Beekman, a combative Yankee in California.

the importation of foreign honey.

"They're dumping it on the market. We ought to go to Washington and shoot some of the worst of the idiots there. That's the way Yankees have always treated anyone who interfered with their rights."

"Beekeepers in Sweden are complaining about honey imports as well," I say. "Including Californian orange-blossom honey that has been heated to maintain its liquid consistency and all its nutritional value has been removed as a result."

"Shoot," says Jack, "you get all those dumb ideas from Germany. They're crazy about their nutritional values over there. We only heat the honey slightly and that doesn't do it any harm. There's been no end of tests to prove that."

I have heard that Californian beekeepers also heat it to stop it fermenting. When you produce honey on an industrial scale and have several thousand hives, it is impossible to check whether the honey in each hive has had time to mature and is ready to be harvested. You just harvest the whole lot which means you are bound to get some immature honey which will ferment unless it is heated. Not that I mention any of this to Jack.

He has already moved on to a different problem, namely the killer bees, a cross between the extremely industrious but aggressive African bee, *Apis mellifera scutellata*, and the docile Italian bee, *Apis mellifera ligustica*, which is the most common race in the U.S.A. In 1957, twenty-six swarms containing African queens escaped from a laboratory in Brazil. They had been imported by a researcher who was trying to breed a better bee than the Italian race, which is not much of a honey producer in a tropical climate. The queens that escaped mated with local drones and the *scutellata* genes proved to be clearly dominant. To make matters worse, the hybrids turned out to be even more easily irritated than their maternal relatives. They were capable of killing humans and cattle and spread to both the south and the north.

"They'll be here soon, too," Jack sighs.

What is to be done? You can't get rid of them with a rifle. There has been talk of setting up nets at the Mexican border, along with corridors filled with insecticide. Would that work, though? In any case he's going on a training course in Louisiana later this year, to learn how to protect yourself against the killers using genetic tricks such as the frequent swapping out of the queens. But that isn't really his style.

His son Bob pops by at this point. He has been to see his mother and is now on his way home with a few pots of orange-blossom honey.

"It has an aroma that knocks you silly!" he says, and then asks what kind of honey I produce in Sweden. He also wants to know how many colonies I have. "Two," I say, and he

It may look a bit untidy inside Jack Beekman's warehouse but he knows where everything is.

probably thinks I mean two thousand.

"And I make a wonderful lime honey that is almost as good as your orange honey."

Father and son look puzzled. "Can you grow limes in Sweden?"

"No," I explain. "Citrus trees that bear limes will only grow in tropical climates, but in temperate ones like Sweden we have a different tree that we also call the lime."*

"So what does that kind of honey taste like?" Bob asks skeptically.

"It knocks you happy," I say, and am gripped by a sudden homesickness for my bees even though it will be many months before the lime trees come into flower. But seeing them gathering pollen from crocuses and snowdrops is wonderful as well.

*I found out subsequently that Americans call the lime or linden tree that grows in the North-eastern states basswood.

Killer Bees for Better or Worse

The invasive Africanised bee – the killer bee – may now have run wild in all of South and Central America, but they are also being farmed on a large scale. They are no trouble if you know how to handle them. On the contrary. They provide huge harvests of honey, are resistant to the Varroa mite and remain unaffected by C.C.D., or Colony Collapse Disorder. It is thanks to killer bees that Brazil and Argentina have become major exporters of honey.

In the U.S.A. they have become established in Florida, New Mexico, Texas, Nevada, Arkansas, Arizona and southern California. Because most American honey farms are run on an industrial scale, they are unable to adapt to the killer bees' temperament as the bee farmers have done further south. Instead they try to block their progress in various ways. The idea of a barrier at the Mexican border has now been discarded, at least as far as bees are concerned. Instead, attempts have been made to insert into their hives new queens that have mated with more placid drones. It is also hoped that as they move northwards they will find the colder climate more difficult to tolerate.

There are, however, beekeepers who have learned how to tame these bees. In "More Than Honey"– a film anyone interested in bees and in their and our future should watch – we meet Fred Terry, who is also well known as a country singer. He sees the Africanised bee as the salvation of the American honey farm, in which more than a third of all colonies die each year. Killer bees are not just more resistant to most of the things that kill other bees; they refuse to accept the mistreatment perpetrated by industrialised honey production. In South America people laugh when they hear that these fantastically productive bees are called "killer bees" in the U.S.A. Why don't they call the cars that kill thousands

The film The Savage Bees *(1978) sowed terror among Americans long before there were killer bees in the U.S.A. Escaped African bees are killing everything in their path and posing a threat to the Mardi Gras festival in New Orleans. In* The Swarm *(1978), Texas has been invaded by killer bees.*

of people every year "killer cars" as well?

"They're not lapdogs, they're wolves!" Fred Terry announces with delight. The techniques for handling them successfully include working on the hives at night or just before first light, and avoiding doing anything to upset them. Their venom is, in fact, no more dangerous than that of other bees, but if they feel threatened the whole swarm will launch a concerted attack and can pursue their target for up to half a kilometer. This means that no matter how good they are at gathering nectar and pollen, they are hardly suited to the amateur beekeeper who wants a couple of hives in the garden for the honey, pollination, and as an integral part of the idyllic life.

Poster by Willard Frederic Elmes, printed in 1929 by Mather & Company in Chicago. The company specialised in printing posters intended to increase the productivity of employees. The industrious bee was an obvious role model. It works for the team and therefore on its own behalf.

APRIL

Various Kinds of Guests at the Hive Remembered
and
A Survey of What People Have Believed We Could Learn From Bees Through the Ages

IN THE HIVES IT'S ALL GO. When the sun is shining the landing boards are just like Heathrow or any other major international airport. Take-offs and landings happening at the same time. According to one young visitor, the coolest bees are the ones that fly straight into the hive with their cargo of nectar and pollen without making a landing on the board.

The more experienced beekeepers who come to visit often say that there is much that bees can teach us humans, although that is more often with reference to the structure of the bee colony than to landing techniques. To quote Maurice Maeterlinck: "the individual is nothing, (her existence conditional only, and herself, for one indifferent moment,) a winged organ of the race. Her whole life is an entire sacrifice to the manifold, everlasting being whereof she forms part."

MAURICE MAETERLINCK (1862–1949)
Belgian writer and poet, awarded the Nobel Prize in Literature in 1911. His 1901 essay *La vie des abeilles* (The Life of Bees) is an impassioned but also instructive account of life in a bee colony that draws many parallels with human behaviour. The Swedish historian Peter Englund wrote this about the work in his blog: "What Maurice Maeterlinck is doing in his book on bees may well be anthropomorphic, not at the level of Disney perhaps, but almost. He ascribes a great many human motivations and feelings to the actions of these insects while arguing with suppressed indignation against those who maintain that their intelligence is only apparent. I think this little book is a marvel even so."

Other visitors know nothing about bees and don't seem interested in knowing anything either – apart from whether they'll get stung – and they cannot be persuaded to go anywhere near the hives. But then there are those who are curious and unafraid. They will ask how bees know where to fly, whether they sleep at night and if they eat their own honey. This makes me happy. The schoolmistress in me loves telling people about life in the hive. But it is rarely the case that I can go on as exhaustively as I would like. More tea has to be made, more juice mixed and conversation shifts to other subjects. Which is why I have put together a fact sheet I can hand out to the inquiring minds among them. Maybe one day they will acquire a hive of their own as a result.

SUMMER IN THE BEEHIVE

At its most numerous, around midsummer, a colony of bees consists of a queen, 60,000–80,000 worker bees and several hundred drones. The queen is the only bee that lays eggs. As many as 3,000 a day!

Worker bees are females who are born from fertilised eggs. They slog away for the whole of their lives. No holidays for them, no raving about self-realisation. They will barely have crept out of the brood cells before they start work: cleaning and caring for the larvae. They are called cleaner bees at this point but within less than a week they are nurse bees. Their hypopharyngeal glands will have developed sufficiently by then to feed "ordinary" larvae with pollen and nectar and the queen larvae with royal jelly, a protein-rich substance they secrete. The special food they receive makes the queens almost twice the size of an ordinary bee and they can live for several years.

The next stage of the worker bee's career is to attend to the cleaning and feeding of her majesty. They also go outside for the first time together with their contemporaries (c. 10 days old) to make orienteering flights. When their glands start producing wax scales they become wax-making bees. They process this substance to construct cells for the brood, for pollen and for honey.

After that it is time to deal with the nectar that their older sisters have brought home in their honey stomachs. Enzymes have to be added and water has to be fanned away if the nectar is to turn into honey. If the temperature in the brood room gets too high they take on the role of fanning bees; if it is too low they expend energy to raise it. They can also stand guard at the entrance and chase off invaders.

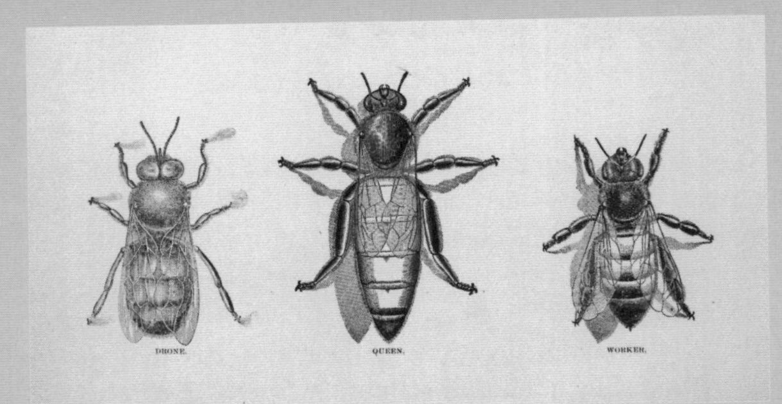

From left: drone, queen, worker bee

Some twenty days after they are born, they go back out into the open as forager or scout bees. In this stage they work harder than ever, collecting nectar, pollen, water and resin from buds that they convert into propolis, an antibiotic putty, that they use to seal cracks. They find out where the nectar-rich flowers are from their sisters, who have already been out scouting the neighbourhood. By performing different round or waggle dances inside the dark of the hive they tell the others in which direction and how far away the sources of nectar can be found, what they smell like and how abundant they are. What is remarkable is that the forager bee will stick to a single kind of flowering plant, which means that the right kind of pistil is fertilised by the right kind of pollen.

But after another one or two weeks the forager bee's wings become worn out and they drop to the ground. A typical industrial injury. Bees born in late summer that spend most of their lives inside the hive may live until the following spring.

In contrast to the industrious lives of the worker bees, the drones – the male bees from unfertilised eggs – appear to lead a life of leisure. They are cleaned and fed and cared for by their sisters. Although they can become very lively every now and then. On sunny afternoons they will fly together with all the other drones in the neighbourhood to a congregation area where they hang around for hours in the hope that a newly hatched virgin queen will show up. If one does, they go completely crazy. The sperm of the ten to fifteen of the fastest and strongest among them that manage to get close enough to permit intimacy will be sufficient to fertilise the half million eggs the queen will lay in the course of her life. Though they have to pay a high price for the pleasure. During the act their sexual organs become stuck and are torn off along with a part of their abdomen, with fatal consequences.

But surely the drones have duties inside the hive apart from sleeping, allowing themselves to be fed and dreaming of mating flights? According to Aristotle (382–322 B.C.E.), "it is good for bees to have drones in the hive because it makes them work harder".

This is true. If the drones are removed, the worker bees lose their drive and more or less cease to carry out their duties. It is said this is because the drones secrete a pheromone that their sisters like. They make the hive feel like a proper home, you might say. They also help to maintain the temperature in the brood area, and there is a theory that they serve as a kind of diplomatic corps too. Unlike worker bees they are actually allowed inside hives that are not their own.

But no matter how much they achieve, with the approach of autumn their careers are over. The time has come for a ruthless massacre when the drones are expelled. The worker bees drive their brothers out of the hive and leave them to their fate, and any that resist are bitten and stung to death. They are not needed during winter, so they cannot be allowed to waste the food supplies stored for the cold season.

The Roman poet Virgil (70–19 B.C.E.*) wrote that bees "were gifted with a spark divine" in a passage on beekeeping in the fourth book of the* Georgics*, an instructional poem about agriculture. He also pointed out that because bees lack any interest in sex, they can devote themselves entirely to work.*

Bees can teach us a great deal – but what?

From antiquity and until very recently bees were likened to exemplary subjects in a perfect monarchy. It was taken for granted that they were ruled by a king because Aristotle had said so in the fourth century B.C.E. and his word – not just about bees but almost everything else as well – would remain the truth for nearly two thousand years.

"Not Ægypt, nor the realms Hydaspes laves / Lydia of vast extent, nor Parthia's slaves, Eye with such awe their King," wrote the Roman poet Virgil (Book IV, *Georgics*). In the view of the Stoic philosopher Seneca, the bee colony was proof that monarchy was ordained by nature and he proclaimed the virtues of the king of the bees to his pupil Nero, the future Caesar. The bee king directed the work of the other bees but – or so Seneca maintained – lacked a sting. "Nature did not wish him to be cruel or to seek a revenge that would be so costly, and so she removed his weapon, and left his anger unarmed. Great kings will find herein a mighty precedent" (Seneca, *De Clementia*). But his message went unheeded. Nero had anyone who displeased him executed, including his own mother. Seneca was ordered to commit suicide, which he did, with Stoic calm even in dying.

The Catholic Church has frequently referred to bees and the bee colony as a paradigm for the faithful. Saint Anthony wrote in the thirteenth century that "Christ Our King flew to us from the beehive, which is the breast of the Father. We should flock to him as bees follow their monarch." In the *Revelations of Saint Birgitta of Sweden*, bees crop up on numerous occasions. The nineteenth chapter of the second book contains the following passage: "I am God, the Creator of all things; I am the owner and the lord of the bees. Out of my ardent love and by my blood I founded my beehive, that is, the Holy Church, in which Christians should be gathered and dwell in unity of faith and mutual love."

But Birgitta also wrote about bees in a manner that is not exactly evocative of Christian precepts. The associations made by people in the fourteenth century were, of course, rather different to those of our sex-fixated age:

> "The Blessed Virgin spoke to the bride, saying: 'Bride of my Son, when you greeted me, you compared me to a beehive. That I certainly was. My body was merely like a piece of wood in my mother's womb before being joined to a soul. After my death my body became again like a piece of wood* when it was separated from the soul, until God raised up my soul along with my body to his divinity. This piece of wood was made into a beehive when that blessed bee, God's Son, came down from heaven and descended into my body. There was prepared in me a most sweet and delicate honeycomb, altogether ready to receive the honey of the Holy Spirit's grace. This honeycomb was filled up when God's Son came to me with power and love and virtue. He came with power, for he was my Lord and God."

The attribute of the bee that the Church was otherwise particularly keen to promote was that it does without the pleasures of the body. The Franciscan monk Bartholomeus Anglicus wrote in the thirteenth century that bees "are not medlied with service of Venus, nother resolved with lechery, nother bruised with sorrow of birth of children". The wax they produced was therefore virginal, like Mary and her son Jesus who was born from her virgin body, and as such perfectly suited for the making of candles for use in church services. The wax symbolised the flesh and blood and Christ, the wick his Soul and the flame his divinity. Medieval theology can

*Before the advent of the straw hive, bees were kept in hollow logs.

At his coronation, Pope Pius XII wears a tiara shaped like a beehive. At the meeting of the International Congress of Beekeepers held in Rome in 1958 he said in his speech to the participants: "We encourage you, dear sons, to see the Lord in the work of the beehive. Praise him with your worship for this reflection of his divine wisdom; praise him for the wax, a symbol of the souls that will burn and be consumed by him; praise him for the honey, which is sweet but less sweet than his words, of which the psalmist sings that they are sweeter than honey."

sometimes resemble psychoanalysis with objects representing something other than they appear to be.

The Renaissance would give new life to the image of the bee colony as a perfect monarchy, with a noble king that had been created by the writers of Antiquity. Erasmus of Rotterdam (1467–1536) had Seneca in mind when he wrote that the lord of the bees is the most powerful of all rulers because he refuses to oppress his subjects and is their benefactor instead. The bees love him and work for him with joy.

Shakespeare described the similarities between the bee colony and human society in Henry V, Act I, Scene 2:

. . . for so work the honey-bees,
Creatures that by a rule in nature teach
The act of order to a peopled kingdom.
They have a king and officers of sorts;
Where some, like magistrates, correct at home,
Others, like merchants, venture trade abroad,
Others, like soldiers, armed in their stings,
Make boot upon the summer's velvet buds,
Which pillage they with merry march bring home
To the tent-royal of their emperor.

Towards the end of the Renaissance the Aristotelian notion that bees were ruled by a king would be challenged, however. First of all by the Spaniard Luis Mendez de Torres who, in 1586, maintained that he had seen *la maestra* laying eggs with

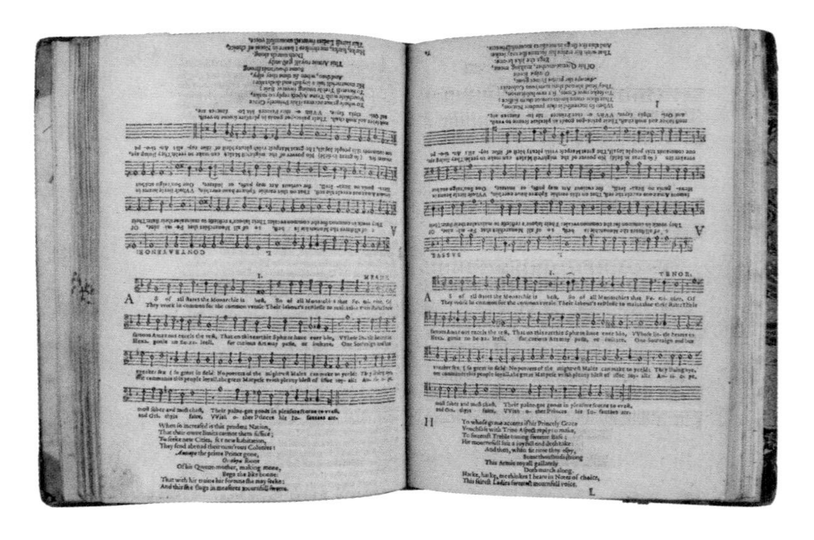

BEE MUSIC

In Charles Butler's bee book, the sounds made in the hive when a new queen is active are recorded in musical notation along with Melissomelos, a madrigal for four voices based on the same bee sounds. It was printed in two directions so that several voices could sing it at the same time.

his own eyes, which meant she could not be male. In 1609 *The Feminine Monarchie* was published by the English priest and scientist Charles Butler in which he wrote that bees had a queen, without, however, providing any evidence for this assertion. But as England had been successfully ruled by Elizabeth I for forty-five years it was not considered improbable that the bees too might have a queen. Butler's book attracted a great deal of attention and is still considered to be a milestone in the literature on bees. Much later the Swede Mårten Triewald would write that he had heard of a book by "an Englishman called Carl Butter" in which he argues that the bees have a queen although "it has become so rare that I have been unable to hunt it down either here or in England." But even without reading "Butter", Triewald was convinced that the queen was female. "Anyone who opens the abdomen of a queen with a flat-edged sewing needle or the thin point of a penknife when she is laying eggs will be able to see the ovaries and eggs for themselves."

JAN SWAMMERDAM (1637–1689)

a Dutch naturalist and the most skilled of the classical microscopists. He was the first person to describe red blood cells and also devoted himself to studying and drawing insect anatomy, including that of the bee.

The first person to study the anatomy of bees using a microscope, and who thus eliminated – or should have eliminated – any doubt about their different genders, was Jan Swammerdam. But despite an increasing body of scientific evidence that proved the "bee king" was not a king, the question remained a sensitive subject throughout the eighteenth century. Human society was steeped in Christian belief, and the bee colony was a divinely ordained model of excellence. What kind of ruler and regime was the right one? Theology and monarchy were ranged against science and the Enlightenment. The debate and discussion of these issues were particularly fervent. In *Spectacle de la Nature* by Noël-Antoine Pluche, an early popular science bestseller from 1732, a count asks a priest who knows about bees whether it is true that bees have a queen. He receives a long and complicated answer – on the one hand, on the other – that concludes with a counter-question: What does the count think? It would have to be a queen, he says, because he has seen eggs in her abdomen. Although, he adds, he has no desire to get mixed up in a disagreement about this with someone who holds a different view.

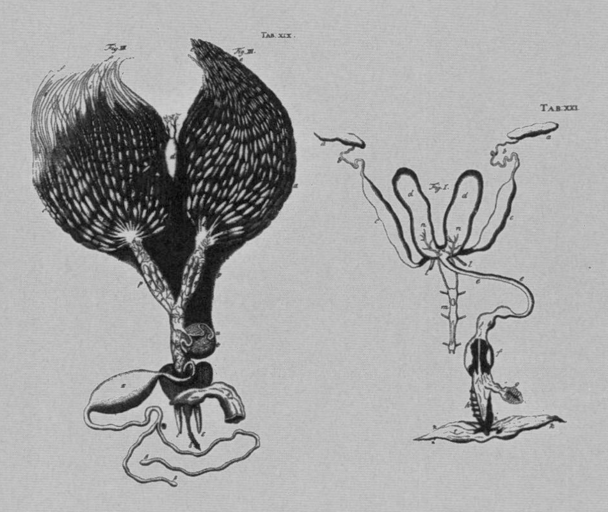

THE SEXUAL ORGANS OF THE BEE

On the left: the queen's sexual organs and on the right those of the drone, as drawn by Jan Swammerdam: "If the reader views the admirable structure of these genital organs, and the exquisite art conspicuous therein, according to their worth and dignity, he will indeed see that God, even in these minute creatures, and their parts, concealed from the incurious eye stupendous miracles," he wrote in his Dissertation on the History of Bees or an Exact Description of their Origins, Reproduction, Gender, Economy, Work and Use.

THE LOVE LIFE OF THE QUEEN BEE

At the end of the eighteenth century, the Slovene Anton Janscha and the Swiss François Huber would discover, quite independently of one another, how the queen bee is fertilised. Prior to this it was believed that the process was not a sexual one. According to Jan Swammerdam, the queen had only to be in the vicinity of the drone whose aura would give life to the eggs she was carrying. Immaculate conception, in other words. Johan Fischerström's account of the love life of bees lies somewhat closer to the truth, but is still very far from accurate:

> "Mating takes place in the following manner: the Female lies on top of a Male and inclines her abdomen towards his, which then projects the previously mentioned double horns as well as the upward bending spike. Once this has occurred several times the Male is dead, which has been observed with a Female when surrounded by one or more Males in particular. These are by nature indifferent so the Female has to employ various proofs of love to inflame their passions. As soon as a Male is released into her vicinity she approaches him, caresses him with her snout, strokes his head with her legs, etc. By the use of such captivating wiles he will eventually be won over and will behave in the end in exactly the same way she does."

Extract from *Nya svenska economiska dictionnairen*
(The New Swedish Economic Dictionary, Part One), 1779

Samuel Linnaeus described how the queen "accompanied by one or two bees, first sticks her head into a pipe (a cell) once her escort has poked at her rear: when she pulls out her head she immediately inserts her abdomen inside which reaches the bottom of the cell and then presumably has laid an egg, while her rear or abdomen remains inserted her escort has been stroking her head as though they were feeding the Queen or caressing her."

And yet despite this observation he hesitates over the sex of the ruler, writing now he, now she. He finds it inconceivable that "a single female can lay 400 to 500 eggs every day from Lady Day to All Saints [. . .] when she has so much else to manage besides." Even stranger is the fact that the queen knows the nature of the egg before it has been laid as he/she lays drone eggs in drone cells, worker eggs in worker cells and queens in queen cells. It is as if "the pups of the shepherd dog had become calves because they were whelped in the cow's byre or the eggs of the sparrowhawk turned into wagtails because they had been laid in the latter's nest."

That is marvellous. One can see the pastor lost in thought as he wanders back and forth within his apiary. What is he supposed to believe? In the end he appears to give up. "Even if the Ruler, who has been called a male in accordance with the traditions of previous ages, may now be found to be a female or Queen in accordance with the information gained in recent times, that is not a matter I think it especially important to know. It is better to behold the intelligent organisation of Nature with a marvellous wonderment and with David say: 'O Lord, how manifold are Thy works.'"

Was that what he really thought? I suspect he had received a warning from a higher religious authority. Questioning the order of the natural world was not acceptable. That could lead to people imagining anything at all. His elder brother Carl (*the* Linnaeus) had received a letter of admonishment from the professor of the Uppsala faculty of theology when he expressed doubt about the fixed nature of living species, which was considered to have been ordained by God.

But the distinguished physician in Uppsala proved to be less of a waverer than his younger sibling when it came to the sex of the ruler of the bees. "See how marvellously the Bee has arranged its Household, how a Ruler or Queen makes loves with so many Drones or Males; she alone has the privilege no other woman has ever enjoyed of the will of men being made subordinate to her own," he announced in his *Dissertation on the Curiosities of Insects* which he held at the Royal Swedish Academy of Sciences in 1739.

Eventually it was no longer possible to question that bee colonies were ruled by a female. But those who found this difficult to swallow might nevertheless devalue her status and appearance. "An awkward creature," wrote the Englishman John Keys, "Not unlike a too tall woman in a dress that is much too short." She was not even a real queen but simply an egg-laying machine.

In the course of the French Revolution, a new set of ideas as to what the beehive represented would come to the fore. It stood first for the collaboration between the nobility, the priesthood and the Third Estate in the

The coup d'etat in Sweden in 1772 increased the power of the country's king, Gustav III, and to commemorate the event he had a medal struck, the Revolution Medal, awarded to those who had supported him. On the obverse it bears an image of His Majesty, on the reverse a beehive and a bee swarm with the inscription "Unanimous in their Loyalty to the King, the Citizens of Stockholm 19th August 1772." It seems to have been particularly difficult for the notion of a bee queen to be accepted in Sweden. The country may still have been haunted by the memory of Queen Christina, the daughter of Gustavus Adolphus, and of her abdication and conversion to Catholicism.

National Assembly. But when the republic was introduced and both noblemen and priests would be guillotined en masse, it became a symbol of diligence and civic spirit. The problem, however, was that a single bee appeared to have power over all the others. The fact that it was a female did not make the situation any more palatable. She was the perfect match for the much-loathed Marie-Antoinette. The status of the bee sovereign – if there was such a thing – had to be determined, and this would be resolved by a debate that took place in 1794 (the year III, according to the Revolutionary calendar) at the new teacher training college, the *Ecole normale*.* It was conducted between a student, Laperruque, and Jean-Marie Daubenton, a seventy-nine-year-old professor and well-known scientist. Previously a member of the Royal French Academy of Sciences and Superintendent of *Le jardin du roi*, the professor had skilfully trimmed his sails to the winds of revolution and would be rewarded with high positions under the new regime as well.

The undergraduate Laperruque began by declaring that when he surveyed the animal kingdom, he discovered there was something worse than a king:

> “I see a queen! And what is more extraordinary, a queen in a Republic. In order to be king, citizen, you said that it is necessary to have courtiers, favourites and favours to dispense, and you added that the lion is not a king because he has none of these [. . .] As for what I am talking about I see around you courtiers, protectors, body-guards, defenders; you see, citizen, that I am speaking of the queen bee. I should therefore wish that natural history should take another step towards republican principles or that it should modify the characters which, according to you, belong to royalty.”

---

*This debate was recorded by a stenographer, as was the case for most of the important debates and meetings of the Revolutionary era.

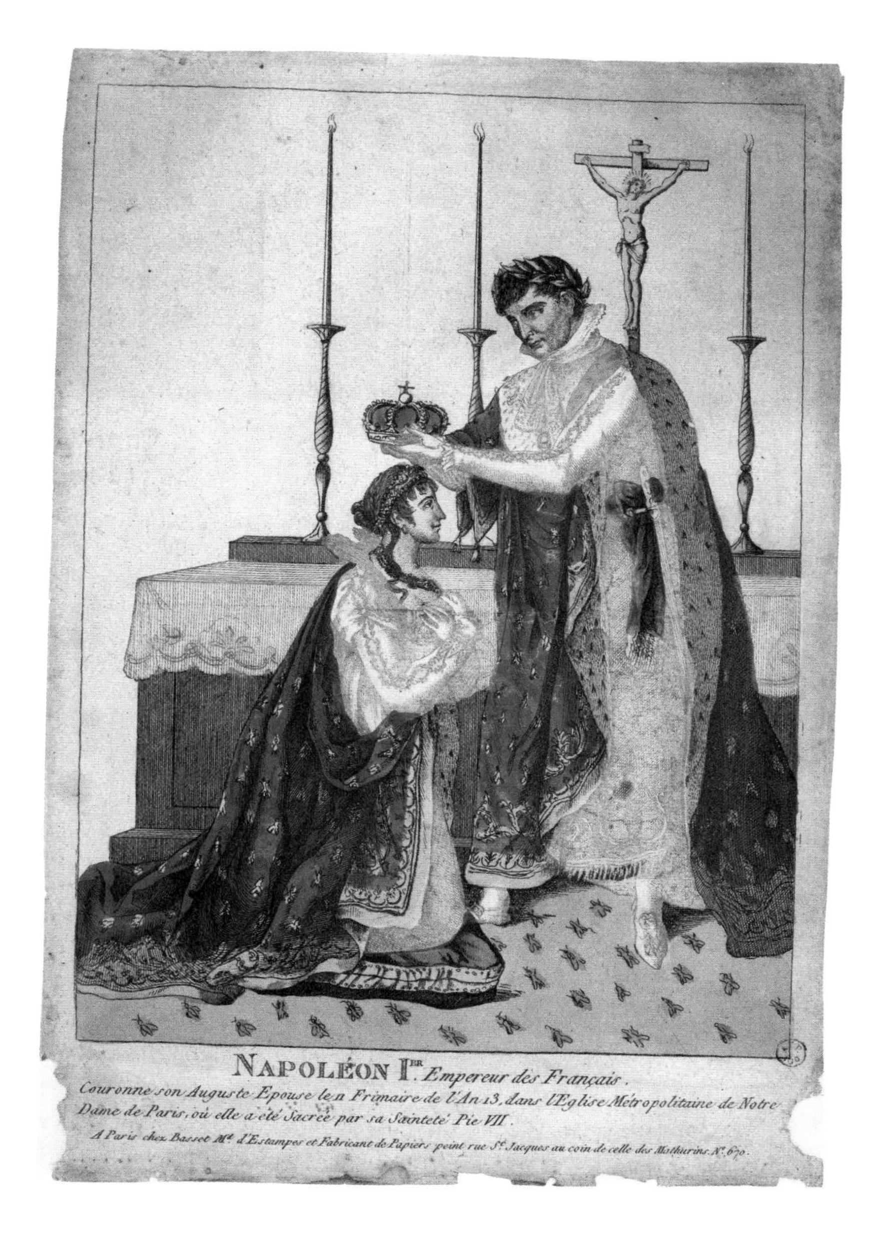

In 1804, Napoleon crowned his wife Joséphine Empress in the presence of the Pope at Notre Dame, after crowning himself Emperor. The couple's mantles are embroidered with bees.

In his response Daubenton called the queen bee *la femelle*, the female, and not *la reine*.

Her only task was to lay eggs and the fact that the worker bees respected her was solely due to the fact that she alone could propagate the species, he said. If *la femelle* should disappear there would be disorder in the colony. Not because the worker bees would receive no orders from above, but because they would be anxious about the continued existence of the colony. They worked for the benefit of the colony of their own free will. The workers were therefore those who ruled.

Bravo! Nowadays we would call that being politically correct.

When Napoleon is forced into exile on Elba he forfeits his imperial attributes. The eagle is flying away, his sword lies broken on the ground and the golden bees are escaping from his mantle. In the foreground is a book of Aesop's fables, opened to a page about a frog that wanted to be as big as an ox. "The stupid beast blew himself up so much he burst."

The metaphorical use of the bee and the colony would really take off at this point. When Napoleon crowned himself Emperor and his wife Joséphine Empress, their red velvet mantles were strewn with bees embroidered in gold thread. Bees were even worked into the carpet they walked on. The bees in question were an allusion to the Merovingian dynasty that founded France. Three hundred bees made of gold, the emblem of the Merovingians, had been discovered in the sixth-century tomb of King Childeric I. These insects, with all their symbolism and wealth of historical allusion, were intended to provide Napoleon's rule with greater legitimacy, and they embellished not only the coronation mantles but the official institutions of the Empire as well as the fabrics, wallpaper and china of the period.

The best known of all bee allegories in history is *The Fable of the Bees* by the Anglo-Dutch philosopher and doctor Bernard Mandeville, a satirical account in elegant rhyming verse accompanied by an explanatory prose text.

The bees in Mandeville's hive bore striking similarities to the England he lived in, the country of nascent capitalism. They steal, cheat, bribe and betray one another; they are vain and self-righteous. Nevertheless the colony thrives and provides well-being for all – until the day a few right-minded bees decide to create an honest society. Out with corrupt officials and wealthy idlers, in with the pure of heart. The only problem is that they have no idea how to get things done. The economy collapses, and because the good bees live only for peace, they have forgotten how to defend themselves against enemies. It all ends in misery, poverty, starvation and death. The conclusion: vanity, hedonism, greed and dishonesty are the incentives required to make the colony thrive. Even criminality is useful. If there were no thieves, locksmiths would starve to death. And what would the Church do if there were no sinners?

The newspapers and periodicals of the day were filled with furious attacks on Mandeville's book. Entire volumes were written about how reprehensible it was. A court in Middlesex declared it a public nuisance. Priests thundered against the impious Mandeville from their pulpits. He was defaming virtue and blaspheming against Christian values. The book caused a furious public debate both at home and abroad. The Dano-Norwegian writer and historian Ludwig Holberg wrote about it, as did Voltaire. Both had objections.

Today Mandeville is considered to be a prominent social philosopher and has been praised by posterity – or parts of it, at least. He has influenced economic theorists such as Adam Smith and John Maynard Keynes as well as neo-Liberals and libertarians such as Friedrich von Hayek, Ludwig von Mises and Milton Friedman. Hayek called Mandeville "a truly great psychologist, if that is not too weak a word for such an expert on human nature". Mises praised him "for having shown that self-interest and the desire for material gain, usually branded as vices, are actually the forces that create prosperity, wealth and civilisation."

BIKUPAN

Tidning för Söndagsskolan och barnen i hemmet.

N:o 17. September 1872.

Till dem som ej ha några gåfvor.

The periodical Bikupan *(The Beehive) was published from 1872 to 1974.*

The worker bee has been the symbol of Manchester since the mid-nineteenth century and is on display throughout the city, from local authority garbage cans and lamp posts to the floor of the town hall. Originally it represented the significance of Manchester for the Industrial Revolution and the hard-working nature of the city's inhabitants, but following the attack on the Manchester Arena in 2017 it came to stand for their unity and solidarity. Many Mancunians had themselves tattooed with bees and the tattooists' fees were given to the families of the victims.

With the rise of the labour and non-conformist movements in the nineteenth century, the issue of who actually ruled the hive would be set aside. Instead, the straw hive came to stand for solidarity, hard work and altruism. The trades union movement in England published the periodical the *Bee-Hive*; in France there was *La Ruche populaire*. In the 1840s, when the Mormons settled in the area that would become the state of Utah, they called it "the Deseret State". Deseret is the honeybee in the Book of Mormon. Utah would then become "the Beehive State", with a beehive as its official emblem. In 1872 the first issue of *Bikupan* (The Beehive), a Sunday school magazine, was published by the Mission Covenant Church of Sweden.

The conservative cartoonist George Cruikshank responded to the labour movement's claiming of the beehive as their symbol with an image of Great Britain borne up by the armed forces and the banking system, with Queen Victoria at the very top. In this hierarchy everyone knew their place and was happy with it.

The perfect monarchy, God's breast, the Catholic Church, chastity, the power of labour, imperialism, social cohesion, thrift, hard work – you might expect that all the possible interpretations of the symbolic meaning of the bee and the bee colony had been exhausted by the time the next century dawned. Something completely new would nonetheless occur in 1911. A single bee, rather than the colony as a whole, would come to serve as an exemplar, heralding a new era with the individual at its heart. She was called Maja and was the creation of Aldemar Bonsels, the author of the children's book *Die Biene Maja* (Maja the Bee).

This bright and curious bee refuses to accept the collective order of the hive and leaves it to live a life of adventure among other insects in the world outside. But when she finds out that some evil hornets are planning an attack on the hive she grew up in, she realises where she really belongs. She returns, saves her people and then stays on as a beloved teacher, sharing her wisdom and experience with the younger generation. An exemplary bee, who is independent and yet profoundly faithful to her origins.

The fact that Maja is a biological impossibility did nothing to stop the book becoming a success. It was translated into forty languages and was published in a special military edition intended to cheer up German soldiers in the trenches of the First World War. Bonsels also wrote travel books and erotic tales; the latter would be burned by the Nazis on their bonfires. Maja survived nonetheless and her popularity even increased during the Hitler era. Despite her wilfulness, the message of the book chimed rather well with the Nazis'

George Cruikshank's image of Great Britain as a well-organised beehive with Queen Victoria and her court at the top and the banking system as its foundation was a riposte to the demands to change society being made by the labour movement.

passion for national folklore and *das Volk*. Bonsels himself was a dedicated anti-Semite and a close friend of some of the highest-ranking members of the Nazi party.

Maja was turned into a film in 1926,* a dramatised documentary with real bees in the main roles. She became a film star once again in the 1970s in an animated Japanese-American production with the more international name of "Maya". Nowadays she is a global star, rather less of a tomboy than in her original guise and also available as a video game, jigsaw puzzle, party balloon, plastic toy and singing doll.

Barry in Jerry Seinfeld's animated *Bee Movie* of 2007 is Maja's brother in spirit, a naughty bee who gradually comes to understand what is important in life. Despite a good education he refuses to work in the hive's honey factory but flies off to see the world, or rather New York. There he becomes close friends with Vanessa Bloome, a florist, and goes with her to a grocery shop where he discovers shelves laden with honey and realises that bees are being robbed. He takes the human race to court and wins the case. The bees are allowed to keep their honey and no longer have to work, which turns out, however, to be terribly boring. To make matters worse, no flowers are being pollinated. Catastrophe is imminent but Barry manages to get nectar collection going again and with it, pollination. Talk about a message! The world needs bees, but also enterprising and daring individuals who dare to challenge the status quo.

Despite Maja's/Maya's and Barry's individualist message, the bee colony as a whole has not yet been exhausted as a paradigm. One new example suggests how a company should be led. *The Wisdom of Bees* (2010) by Michael O'Malley, a beekeeper and professor at Columbia Business School,

*An extract from this extraordinary film can be seen on Youtube: *Die Biene Maja und ihre Abenteuer*.

focuses on decentralised decision making, long-term planning, job rotation, clarity, functioning feedback and set routines as the inspirational keys to the success of bees. Most significant, however, is the fact that everything *Apis mellifera* does is for the benefit of the hive. This mindset is essential to the success of any company.

In *Survival of the Hive* (2013) by leadership experts Deborah Mackin and Matthew Harrington, the hive is presented in a somewhat jokey way as a living illustration of the importance of confidence and incisive leadership to the success of a company. When faced with any kind of decision, managers should be asking themselves whether it will favour the long-term survival of the organisation.

While the bee colony has become a symbol for companies whose goal is profit, it has also come to signify the very opposite: for an ecosystem badly damaged by human greed. In the past, miners would take a canary with them below ground; its death would suggest the presence of noxious gases. The bee is our modern-day canary.

Maya the Bee has become an international celebrity. Her original name was Maja and she was the heroine of Die Biene Maja und ihre Abenteuer *(The Adventures of Maja the Bee) by Waldemar Bonsels, published in 1912.*

But how does the bee colony *really* work? One of the most recent accounts is to be found in the work of Thomas D. Seeley, a prominent American bee scientist who has studied decision-making in a swarm choosing a new home. His conclusion is that this process takes place democratically – and that there is much we could learn from bees. Does that sound familiar?

In *Honeybee Democracy* (2010) he describes what happens when a swarm has left its old hive and opted temporarily for the branch of a tree, say. Several hundred of the most experienced scout bees form a reconnaissance committee that flies off for several kilometers, up to five thousand meters, in every direction to inspect the hollow places they find on the way: chimneys, old tree trunks, ventilation ducts, bird-nesting boxes, rocky caves. When a scout finds a space that appears attractive, she inspects it for at least forty minutes. When she returns to the swarm she performs a dance on the backs of other scout bees to inform them where the new site is and what it is like. Other scouts describe other possible sites. Those which appear particularly promising are visited by yet more bees, which then do their own dances, and comparisons are made. The selection process takes several hours, sometimes even days, but for all practical purposes invariably ends with the swarm moving into what is the most suitable hollow site.

Seeley describes how bees make decisions by sharing common goals. While they contribute different skills, they have the same preferences and freely share the knowledge they have obtained with one another. No pressure is exerted and the decision is made by a qualified majority. Although few groups of human decision-makers can function as optimally, this must be a good thing to strive for. This is the goal Thomas D. Seeley has set for himself as head of the Department of Neurobiology and Behavior at Cornell University, and he thinks it works fairly well.

Cover to the sheet music for "Never Let the Same Bee Sting You Twice", originally a song by the blues guitarist and composer Richard "Rabbit" Brown. Isn't that an odd title? After all, when a bee stings you the sting is torn off together with part of the abdomen and the bee dies and can never sting again. Brown was not singing about real bees, however, but about loose women. Then Cecil Mack and Chris Smith stole the title for their own song, as the indie band The Veils would later too.

MAY

A Visit to Lennart and His Girls Remembered

and

Answers to the Questions: Why Do Bees Sting?, How Are Bee Stings Best Treated? and Is Protective Clothing Necessary?

Popping in to see Lennart Kuylenstierna is always fun. He lives on my street and keeps bees too. The last time I visited he was in the garden cutting the grass, neatly dressed as always in a jacket, shirt, tie and linen trousers. He is a gentleman. A Lennart in shirtsleeves is as unthinkable as a Lennart in a tracksuit, or jeans or shorts. Or in a bee suit for that matter. A bee hat is as far as he will go when it comes to protective clothing.

He was wearing the hat when I arrived and the veil had been lowered.

"Don't go too close to the hives," he called over. "My girls are in a really dreadful mood today."

He had just been checking on his hives and removed some queen cells before he started making a din with the lawnmower. The bees had got very annoyed and were circling him in threatening fashion. Lennart did not turn a hair,

Lennart Kuylenstierna and his rather splendid apiary.

however – he just got on with the mowing. It needed doing, after all.

He was given his three hives by the Dominican nuns in Rögle, outside Lund. They had experimented with beekeeping but gave up after a year or two. He painted the hives honey yellow with white gables and fitted them with name plates: BIFROST, VISINGSBORG and DROTTNINGHOLM. Plants that bees love bloom around his hives from early spring to late in the autumn: sallow, winter aconite, crocus, scilla, broom, fruit trees, raspberries, lavender, oregano, thyme, fennel, borage, marigolds, buddleia, sunflowers, windflowers and ivy. There are very few bees in Lund as well looked after as Lennart's.

Once he had turned off the mower, we sat in the arbour to discuss bee matters. But as soon as he lifted his veil a bee came along and stung him right on the forehead.

"Ulla-Stina, how silly of you," he sighed.

Which, of course, it was. The sting of a bee that has stung someone or something is torn off along with the venom sac, and the bee dies. Goodbye Ulla-Stina. I held up my pocket mirror so he could see to pick out the sting. Then he fetched an onion and cut it in two with his Swiss penknife – the kind a gentleman always has on his person – and rubbed the swelling with it.

"Onions are the best remedy against bee stings," he said.

Then he pointed over to the hives where the young bees were practising their buzzing technique with wings that glittered in the sunlight.

"Look at my girls – they're gorgeous, aren't they?" he said.

Lennart loves his bees. Getting stung every now and then matters not at all.

Why and whom do bees sting?

The ancient Romans also used raw onion to soothe the pain of bee stings. Other remedies recommended in the course of history have included chopped parsley, honey, hot cow dung diluted with vinegar, mashed parsnip leaves, chewed mallow, breast milk whisked with egg white and rose water, tobacco, toothpaste, wet clay, your own urine, a little boy's urine, quicklime, scorpion oil and lumps of ice.

When you have been working with bees for a while you notice that some people almost never get stung, while others cannot go near a hive without being attacked by furious bees. Even moving calmly and cautiously, one of the ground rules you have to learn as a beekeeper, proves to be of no help to them.

And what about their sense of smell? Aristotle observed that bees are made aggressive not only by unpleasant smells but by perfume as well. According to the Roman writer Columella, anyone approaching a hive should not be drunk, have eaten garlic or have failed to wash. You were not supposed to have had intercourse recently either. Pliny's blacklist was similar, but he added that bees loathe women who are menstruating.

The bit about intercourse and menstruating women does sound a bit odd. The ancient Romans were not moralists though, and people were bound to have smelled more strongly, particularly at certain times, before deodorants and daily showers were taken for granted. And especially for a bee whose sense of smell is a hundred times more powerful than a human being's.

The Christian theologians who wrote about beekeeping a millennium or so later most definitely were moralists, however. They were also well read in the work of the classical

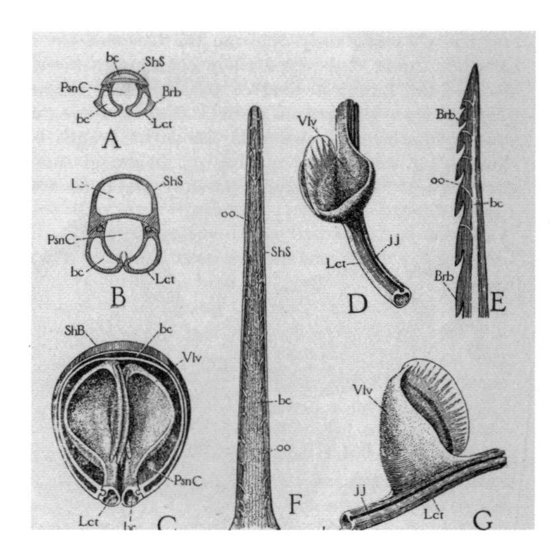

Anatomical studies of the various parts of a worker bee's complex venom apparatus. From The Anatomy of the Honey Bee *by Robert E. Snodgrass. Of the stinger itself, P. Gullander and J. O. Hagström wrote in 1773 that "it is pointed, thicker above than below, and is split at the tip and hollow along its length so that the venomous liquid can flow along it."*

Woodcut from Olaus Magnus' History of the Nordic Peoples *of 1555, showing how bees respond when intoxicated people with foul breath, "their greatest fear", approach their hives. They attack! The couple on the bench may also eventually find themselves at risk as according to Olaus Magnus bees also hate people who have just had intercourse.*

writers of Antiquity and appear to have relied more heavily on them than on their own observations. In addition, they were pious men who saw God's will in everything, including bee stings. "But if thou will have the favour of thy bees that they sting thee not, thou must avoid such things as offend them, thou must not be unchaste or uncleanly, for impurity and sluttishness (themselves being most chaste and neat) they utterly abhor," wrote Charles Butler, the vicar who has been called the father of English beekeeping. The Swedish bishop Olaus Magnus would go even further in his *Historia om de nordiska folken* (History of the Nordic Peoples):

> "Never has a battle as bitter been waged against so hostile an enemy as when bees point their stingers at an intoxicated person with stinking breath – their greatest fear – should he approach their dwellings. (....) Thieves, pimps and menstruating women have a particularly adverse effect on bees if they come near their hives.

The same is true of those who have had intercourse and those who have consumed acid and foul-smelling food as well as salted food of all kinds. ”

Unlike Olaus Magnus, who seems to have largely plagiarised antique authors, Samuel Linnaeus based what he wrote on his own experience. Linnaeus' advice to anyone who had acquired a new bee colony was to breathe through the entrance to the hive every evening so the bees are comfortable with the keeper's smell. He also maintained that "bees know the person who attends to them daily and prefer not to sting them even if they smoke tobacco and drink brandy."

Presumably the parson himself had both smoked tobacco and drunk spirits before visiting his hives. Could it be the unsteady motion of the drunk that irritates the bees as much as the smell of alcohol? For my part I have met beekeepers, like Linnaeus, who smoke and drink without arousing the rage of the bees, but then they would also remain perfectly calm when handling the insects.

Linnaeus continues: "They do not readily tolerate newlyweds, further acquaintance makes no difference, nor women at certain times." Hmm. "Newlyweds" must be a euphemism for what we would call sex nowadays, which beekeepers have been warned against ever since classical times, just as they were against menstruating women. Did Linnaeus also base this statement on personal observation? Or was he simply passing on established prejudices? We can only speculate. Or make observations of our own.

The French zoologist René de Réaumur, on the other hand, dismissed all the old theories about what induced bees to sting. The ones about sex and menstruation were rubbish as well:

"If certain authors are to be credited, one should not approach bees without having first examined one's conscience. Those writers assure us that bees will not put up with dirty individuals and particularly not with those who are guilty of adultery, and that they will show no mercy towards thieves. They are virtuous insects who love those who are virtuous and are capable of distinguishing them from the depraved. It has also been claimed that there are times when ladies should not approach them. But all these putative aversions in bees are merely tales and inventions."

RENÉ DE RÉAUMUR (1683–1757)
One of the most brilliant scientists of the Enlightenment. His works include a vast treatise on the history of insects, *Mémoires pour servir à l'histoire des insectes*, which deals with the honeybee in the fifth part. He was able to study its life in detail thanks to the glass beehives he had invented.

The interest shown by writers on bees in the way people smelled would wane over time as hygiene improved and those in professions other than theology began to write about beekeeping, but it is still mentioned from time to time, as in *The Urban Beekeeper* by Steve Benbow (2012):

"Many people don't realise it, but bees have an impeccable sense of smell. Just as we do, they react badly to unpleasant odours; it makes them stressed, grumpy and more likely to sting their handler. To get on well with them you need to be clean and odour-free."

The earliest protective clothing for beekeepers in Europe was a kind of hooded cape with some transparent fabric or netting in front of the face. The most frequently used piece of equipment was, however, a hat with a veil. Today most beekeepers in Sweden keep friendly Buckfast bees; even so, they usually wear all-in-one bee suits that make them look like astronauts.

Nowadays if writers on bees mention smells they also warn against shampoo and perfume in addition to alcohol.

Eva Crane, that most expert of bee historians, mentions that special protective clothing for beekeepers has existed since the fifteenth century, even if not much was needed at that time. Everyday clothing, both men's and women's, covered most of the body and all that was required was a hood fitted with netting that could protect the face, plus gloves when necessary. And yet in Pieter Bruegel's intriguing drawing of 1568, the beekeepers – or rather bee-thieves – are kitted out from head to toe and have wicker screens to protect their faces.

EVA CRANE (1912–2007)

An Englishwoman with a doctorate in atomic physics, she eventually switched to studying bees and beekeeping. She visited sixty or so countries to gather material for her books and articles. Her major works are *Bees and Beekeeping* (1990) and *The World History of Beekeeping and Honey Hunting* (1999).

"Although we consider a bee sting not worth mentioning, we nevertheless recommend for practical reasons a straw hat with a cloth veil, which can be tucked inside one's coat. Wearing a uniform to approach the bees, however, is ridiculous and no-one with any self-confidence would do so."

This condescending quotation is from *Stora Biboken* (The Big Bee Book, 1909). Nowadays the uniform – a bee suit – can actually provide a feeling of self-confidence by helping to define the wearer as a real beekeeper.

Peter Vingesköld, a beekeeper on the Swedish island of Gotland, is one of those lucky individuals who almost never gets stung, even if he has drunk alcohol, and he never needs to wear protective clothing. When he was due to be photographed for a newspaper article, however, the photographer protested. How could you tell Peter was a beekeeper if he wasn't wearing a white overall, gloves and a hat with a veil? It made no difference that he was standing among the hives with bees buzzing all around him. In the end he was photographed in his ordinary clothes, but at some considerable distance from the hives.

WHAT ARE THEY DOING?

Pieter Bruegel the Elder's intriguing drawing of the 1560s, showing three men wearing beekeeping clothes and a figure in ordinary dress up a tree, has been interpreted in many ways. The picture is clearly allegorical rather than a practical depiction of beekeeping. What are those three suspicious figures really up to? One theory is that the picture is about the conflict between the Spanish Inquisition and Flemish Protestantism, and may be meant to show beehives – symbols of the Catholic Church – being stolen and vandalised by wicked Protestants. Bruegel was a Catholic. But it may also mean the exact opposite: tacit criticism of the Inquisition.

Look, bee swarms! Illustration of Virgil's Georgics *by Wenceslaus Hollar, a seventeenth-century Bohemian engraver.*

JUNE

A Swarm that Caused Problems Remembered

and

Accounts of Rows between Neighbours and How People Dealt with Swarms in the Past

THE SWARMING SEASON IS UPON US, and now it is all about being prepared. When a new queen is hatched, the old one flies off with thousands of her loyal retainers to establish a new colony. It is an extraordinary sight, and what a noise they make!

> "They hurl themselves out of the entrance as though blown by a hurricane, then they fly back and forth in the air with a loud buzzing sound only to land in a tight cluster around a nearby spot, where, if the queen is with them, they stay still for quite a while."
>
> From *Nordisk familjebok*, a Swedish family encyclopedia

Swarming is both a primal force and the way in which a bee colony reproduces. If you want to increase the size of your apiary you need to have a new hive ready in advance, so that when you capture the escapees you can settle them in their new residence. In the past the inside of the hive would have been rubbed with fragrant herbs to make the bees feel welcome. A lovely custom and a good tip for today's beekeepers.

If, however, the swarm manages to get away, you will not only lose a few tens of thousands of worker bees but also the equivalent of a couple of jars of honey that they take with them as food for the journey. Which is why I have learned to make regular checks on my colonies from early spring until well into July, to make sure they have enough space. I lift out the brood frames one by one to inspect them, and if I spot a cell that is bigger and different to the others, it contains a queen-to-be. I take out my hive knife and get rid of it.

But once the bees have made up their minds to swarm, it is difficult to stop them. Despite fairly regular inspections they will swarm every season, and almost always at inconvenient times. Last year a neighbour rang just as I had got back from Stockholm and only an hour before I was due to set off for the dentist.

"Get over here, now. There's a swarm of your bees in my pear tree – you need to deal with it!" he said.

I went over and had a look.

"Those aren't my bees," I lied. "They're Lennart's."

Lennart takes great pains to inhibit any tendency on the part of his bees to swarm, and the idea that it might be his girls in the neighbour's garden was actually pretty far-fetched. But the lads who usually help me capture swarms were away. And, as I said, there was the dentist.

The neighbour looked skeptical. He is not particularly well disposed towards bees, at least not to mine, even though he should be grateful: after all, they pollinate his fruit trees and berry bushes.

He also complains that they drink from his bird bath. Seeing them on the surface of the water isn't very nice, he says.

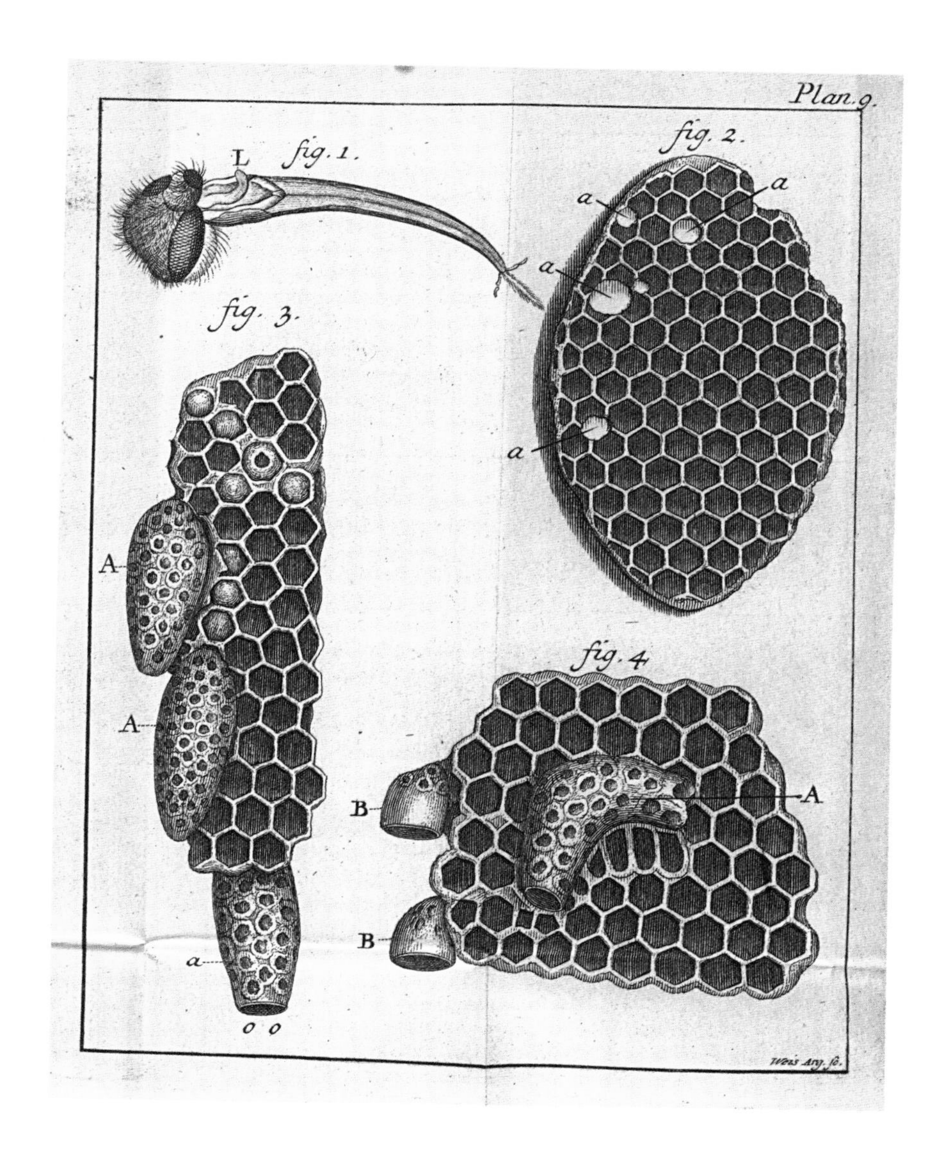

On the two lower sections of the wax combs you can see the oblong queen cells that are clearly different from the capped worker-bee cells – the lighter ones at the top of the cross-section containing the three queen cells. Illustration from Histoire naturelle des abeilles *(Natural History of Bees) by Gilles Augustin Bazin, 1744.*

My suggestion of putting up a sign with bees crossed out leaves him unamused.

He is critical in general of how I manage my garden and his accusations include seeds from the veronica in my lawn flying over to his. He considers it a weed and has told me I need to spray the lawn. I think it looks wonderful when the veronica turns the lawn a light shade of blue, and I won't have anything to do with poisons.

A classic conflict between neighbours. For one it is about the beauty of diligence and order set against laxity and lack of care, for the other, life and lushness as against discipline and control. The neighbour and I are civilised human beings and we are able, despite our ideological differences, to interact socially and even have a fairly nice time together. Particularly when we've had a glass or two. But sometimes things blow up between us, as they are doing now.

Fortunately, the swarm headed off southwards before he could ring Lennart and discover that his bees had not swarmed at all. I felt relieved, but I was also a little ashamed. A good beekeeper stays home during the swarming season and keeps watch for when the bees are getting ready to leave. A swarm on the loose in a densely populated area can really disrupt the lives of people who are entirely uninvolved. Eventually it will settle somewhere permanently, and if that is in a ventilation shaft or beneath roof tiles it can cause a lot of problems.

Touchy neighbours have always been with us

Even in Ancient Rome spats took place between neighbours over bees. There is an account in *Aper pauperis* by the Roman rhetorician and lawyer Quintilian (35–100 C.E.) – or one of his pupils – of a peculiar legal case.

A poor man and a rich man were neighbours in the countryside. The rich man had flowers in his garden, the poor man had bees. The rich man complained that his flowers were being damaged by the visits of the pauper's bees and demanded the latter be moved. Because the poor man did not move them, the rich man sprayed poison on his flowers. All the poor man's bees died. The rich man was now being sued for the damage he had caused.

Swarming bees do not attack people under normal circumstances, but if you start hysterically waving your arms around, yelling and running around like the picnickers in this image are doing, the bees feels threatened and may become aggressive. Aquatint by George Cruikshank, 1826.

How preposterous this seems nowadays when we know about pollination. How on earth could anyone think bees could damage flowers? But humility is vital in this regard. Much of what the courts deal with in our time will seem completely incomprehensible two thousand years from now.

To judge from the newspapers, today's conflicts over bees tend to involve neighbours who are afraid of being stung – apart, that is, from the ones about bee poo.

“The neighbour of a beekeeper whose hives are kept at the corner of Frejgatan/Långgatan has complained to the environment committee. The neighbour says he has been stung by the beekeeper's bees. The matter has now been discussed by the environment committee who decided that the neighbour lives much too far from the hives for environmental protection law to be applied in order to prohibit the keeping of the bees.

'We cannot be certain that it really is this keeper's bees that did the stinging. The committee also considers that the benefits provided by the bees outweigh any harm. Without bees there would be no flora, no vegetation,' says environment affairs director Lennart Rydberg.”

Helsingborgs Dagblad, August 5, 2003

Now that is an intelligent civil servant. Besides, it was probably not bees that did the stinging but wasps. Forager bees are kept fully occupied gathering nectar and pollen and have no reason to attack people unless they feel threatened.

“Last March a homeowner in Everöd installed two hives on his plot of land. He had checked on the rules for beekeeping in advance and discovered that there are no clear legal provisions on the matter. Because he considered that the area was sufficiently open, he did not think having beehives on his land would cause any problems. The neighbours, however, thought otherwise.

More or less immediately after the beehives had been put in place the neighbours sent in a complaint to the local office for environment and health protection. The homeowner was informed that the neighbours had complained and was also told that 'active monitoring' might be implemented and that the committee would decide if the complaint was justified. And in that case he would be billed for the monitoring. The homeowner still thought that he had done nothing wrong and refused to remove the beehives. Active monitoring was introduced and after various twists and turns it was determined that the bees were a disturbance to the neighbours. The area was considered too densely inhabited and the neighbours maintained they were allergic to bees. The homeowner then removed his hives. A few weeks later he was a sent a bill by the environment and public health committee for SEK 3,200 for the cost of the monitoring."

Kristianstadbladet, July 28, 2012

How shameful!

Bees can be the source of legal conflicts between neighbours too, as in the case of Kearry v. Pattinson (1939) found in the *Cambridge Law Journal*:

"Some of the bees of K., a bee-keeper, swarmed and settled in the garden of P., his neighbour. P. at first refused K.'s request for permission to enter his land in order to rehive them, but after some interval he gave it. K. entered P.'s garden, but found that the bees had flown. He thereupon brought an action against P., claiming that P. had prevented him from entering P.s' land, or alternatively refused to permit him to do so, with that result that his property had been lost, and asking for damages for loss of the bees and of the profit which he would have made on their honey. His claim was based mainly on conversion. The county court judge found as facts that K. had not lost sight of the bees and could identify them, and held therefore that they remained K.'s property. "

After an appeal by K., the case went on to establish that bees are animals *ferae naturae*, and that once they had settled on P.'s land, K. ceased to own them. K., or indeed any other trespasser could have gained ownership of the bees by hiving them, and any interference by P. would not have rendered him liable.

In the past a bee swarm could be a much more precious commodity than it is today. The Swedish Country Law of Christopher dating from 1442 stipulated that the theft of bees was punishable by hanging, and medieval Scandinavian law laid down in detail who owned the bees: whether their finder or the owner of the land. However odd it may sound, the rule awarding the finder one third of the bees with the rest going to the landowner contained in chapter 21 of the Byggningabalk (Swedish statutes concerning agricultural law) of 1736 still applies today.

" If one finds bees in one's own dwelling place or in one in which one has a portion, the bees belong to the finder. If in the countryside, one third goes to the finder and two thirds to the landowner. If one finds bees on someone else's enclosure one has no entitlement; in someone else's woods or grounds one third while two shares go to the landowner. If two men claim to have found the bees, the reward goes to the man who was the first to claim them. Provisions concerning the person who finds bees on someone else's land and removes them but makes no claim, and concerning the person who lures to himself another man's bees with food and bait are laid down in the *missgärningabalken* (the Swedish Penal Code of 1734). "

But when the ethnologist Albert Sandklef carried out research into the subject in the early 1940s, none of those interviewed

Everyone out of the house! As much noise as possible had to be made when bees started to swarm. Sixteenth-century engraving by Jan van der Staet.

were familiar with the law as it stands. "I have asked the question of bee-owners in the Swedish provinces of Scania, Halland, Småland and Västergötland and all the people questioned gave identical answers: the landowner and the finder each get half." Nowadays no landowner is likely to lay claim to found bees. They are more than grateful if someone is willing to look after them.

In the past it was customary to make a lot of noise when a swarm took off. People banged saucepan lids, rang hand bells, clapped their hands and hooted. It was believed that this made the swarm clump together more rapidly. Why though? The Roman writer Columella thought that bees were frightened by the noise, but according to Pliny they liked it and it helped calm them down. Pliny's explanation is the one that has survived longer.

"When you see them beginning to emerge in large numbers, pick up cymbals, a hand bell, some other bell or instrument and ring it slowly under a juniper branch, because the lovely sound makes them cluster together and settle down," wrote Isaac Erici in the mid-seventeenth century. In the recorded traditions of both Denmark and Sweden there are accounts of bees gathering together to enjoy the sound if a hand bell was rung or an instrument was played.

During the Enlightenment, when most matters began to be considered from a more skeptical point of view, this notion would be challenged too. "The kind of hellish noise they make in the country by banging on saucepans and copper vessels tends rather to scare the bees away than encourage them to gather together as bees do not like noise. It is not known when this ridiculous and superstitious custom arose, but it was already prevalent in Antiquity," according to Diderot's *Encyclopédie*. Samuel Linnaeus maintained that no ringing of bells, hand bells or other such things should take place when bees swarm. The important thing was to make room for the swarm and not get in its way.

Not many beekeepers read books in the eighteenth century, however. You learned the skills required from older people, and it took a long time for customs to change. Well into the twentieth century, people were still ringing bells or banging saucepan lids when bees began to swarm. The ethnographer Carl-Gustaf Bernhardsson recounts how on Skaftö in the 1930s, "a man would ring a bell to make a flying swarm settle and not fly away."

A third explanation is provided by the insightful Hans Herwigk. The noise was a means of informing the neighbouring area that a swarm was in flight. "It is often necessary to beat a pan when a swarm is in the air because there may be several people and not just one who keep bees in a particular neighbourhood. If the bees fly into someone else's garden and the owner has banged on a pan, the other man cannot then say that the bees belonged to him."

That does sound reasonable. People can hear while bees lack organs for hearing. On the other hand they are extremely sensitive to vibrations, and there would have been a great deal of vibrating when people banged metal objects and rang bells.

This process is known in English as "tanging", and there are many eyewitness accounts on the Internet – mostly from American alternative beekeepers – that it works. A swarm will cluster even when all you have is a plastic bucket and spade to bang together. You can apparently even get a swarm to return to the hive through tanging. So what are we supposed to think? We are always free to experiment.

The shopkeeper in Hawnby showing his hives to Karin. These are based on the iconic British W.B.C. model of 1890, impractical and difficult to put together, but beloved for its appearance. Though some call it a Victorian monstrosity. How typically English.

JULY

A Walk in Yorkshire Remembered

and

A Comparison between Heather and Chestnut Honey

I AM WALKING ALONG the Cleveland Way in Yorkshire with my Norwegian-born friend Karin Nihlén. We pass through green wooded valleys and through villages with houses covered in climbing roses before moving up onto the moors where the sheep are bleating and the lapwings wail.

This is a wonderful area, but it rains a good deal and fog settles every now and then. Is the weather the same at home? If it is raining in Lund now too, there may be much less lime blossom honey. I'll have to ring home to check.

Karin, who doesn't have bees to worry about, is dreaming about the sun-drenched Italy she loves, where she is thinking of buying a house with money she inherited from her mother. She is not an Anglophile like me, and going walking in Yorkshire is mostly an opportunity to think through whether to buy that house or not. She never stops comparing England with Italy, always in Italy's favour.

I respond with examples of how unhelpful the Italians can be – unlike the English. I find that unless you know the language it is hard to get by in Italy. And speaking as a journalist, my experiences of that country have not left me favourably disposed towards it. Karin, who is fluent in Italian, replies that they know how to make the most wonderful food, unlike some other nations. We continue bickering in this vein, it's great fun.

After lunch on the third day the fog has become so thick that we cannot find the trail. Having wandered around on paths the sheep have trampled and that are not on the map, we arrive in the little village of Hawnby. We ask at the grocer's shop whether there is somewhere we can spend the night.

"You can stay here," the shopkeeper's wife says. "We've got an empty room in the attic. And my husband will help you up the stairs."

He appears from the rear of the shop. We introduce ourselves and Mr and Mrs Banks ask where we are from.

"From Norway," Karin says, even though she has lived in Sweden for thirty years.

I nod in agreement. I've been to England before with Norwegian friends and noticed how warmly they are received. As it happens the British are usually friendly to tourists, but especially so when they come from Norway. This has something to do with the Second World War. It is still fixed in the collective memory somehow that the Norwegians mounted a particularly brave resistance to the German occupation. The Swedes, on the other hand, were weak and allowed German troop transports to pass through their country on their way to and from Norway. In 1945 Churchill wrote a memo to the British chiefs of staff about "the calculated selfishness that has characterised the Swedes during both wars against Germany."

No bees can be glimpsed in the Yorkshire fog.

It's not going to make any real difference if our hosts think I'm from Norway as well, is it?

We change into dry clothing and when we come back down again, tea and scones have been laid out in front of a cozy coal fire. What a wonderful country England is. I notice how comfortable Karin appears to be in her armchair, but I don't comment on it out loud.

"You ought to come here in August," Mrs Banks says when we thank her for the rations. "It doesn't rain nearly so much then, and it's gorgeous when the heather is in bloom."

"Heather!" and immediately I'm lost in thought about the bees again. Heather honey is amber yellow and tastes exquisite. "Is there a beekeeper in the area we could visit?"

"Of course!" she says. "My husband's just having a look at his hives now, on the other side of the road."

Karin tags along, even though bees hold no interest for her. She doesn't mind seeing the shopkeeper again. A charming man, as both of us immediately noticed.

It's all go at the hive entrances. There's a lot of foraging to do despite the weather. The white clover is in bloom and it needs moisture if it is to release its nectar, Mr Banks explains.

"Clover honey is delicious," he says, "but not as delicious as heather honey. I sell it on the comb and people come all the way here from York to buy it."

But when I get out the camera he really begins to put on a show. He shifts his bee hat so it sits askew, gives Karin the smoker and lifts the roof of one of the hives.

"Give them some smoke."

She is appalled. Surely this madman isn't going to unleash his bees on her?

He doesn't, of course. Beekeepers are almost always nice and friendly, one-to-one at least. Though I've never met such a chirpy chappy as the grocer in Hawnby before, nor anyone whose hives so sorely needed a coat of paint. People in England are not as obsessed with keeping everything spick and span the way we are in Sweden, but on the other hand there is a cosiness and a warmth here that you rarely find at home.

A real English breakfast is served the next morning, with all the trimmings. But Karin wants espresso and cornetti. She refuses to eat eggs, bacon, fried mushrooms, fried tomatoes and chipolatas. In order to avoid upsetting Mrs Banks she forces me to eat most of what is on her plate as well. Meanwhile Karin munches on a slice of toast and strengthens what she calls the absolutely useless coffee with several teaspoons of the espresso powder she brought with her.

Eventually we get back on our walk. The fog has lifted and the sun is peeking through. Suddenly Karin says she has finished thinking it over. There will be a house in Italy. This is something we have to celebrate and in the next village we come to we go into the pub and order a glass of white wine each. It tastes acidic and peculiar. Completely undrinkable. Karin sighs. These English.

"Where does this wine come from?" she asks the girl behind the counter with barely concealed irritation. The barmaid is surely more used to drawing pints for the locals than serving wine to foreign ladies in hiking boots and carrying rucksacks.

"It comes from a bottle," she replies in astonishment.

Karin groans. You don't get stupid answers like that in Italian trattorias.

"And you can get chestnut honey there that tastes much better than your heather honey," Karin says.

Are you a heather person or a chestnut person?

I have now tasted Italian chestnut honey – from sweet chestnuts, not the horse variety – which makes me feel even more sad that Karin is no longer alive. She would have been so pleased to hear that I think it has an unusually lovely taste. It is spicy, slightly bitter and definitely not keen to please. Like Karin, I curl it over a piece of *pecorino toscano* and miss her.

I have learned from the Internet that chestnut honey – which can also be found in France, Spain, Greece and other southern countries – can counteract ageing due to its high antioxidant content; it relieves tiredness, cures acne, strengthens muscles and the immune defence, improves blood circulation and digestion and so on.

But, if we are to believe the Internet, heather honey is no less remarkable. It helps against rheumatism and frostbite, urinary infections, anaemia, exhaustion, menstrual cramps, poor kidney function and osteoporosis. A more muted though still positive assessment can be found in the peer-reviewed study carried out by the Karolinska Institute, one of the world's leading medical research centres, in 2006. It found that heather honey also has superb antibacterial properties. Not everyone appreciates it, however.

"If the flower is bitter, the honey will be too. Heather and buckwheat will always produce a honey that is harsh, and no honey with a good flavour can therefore be produced where these plants exist in large quantities," according to a Royal Swedish Academy of Sciences report from 1773. "Heather honey is brownish in colour and is far inferior in terms of taste and value to ordinary honey," wrote the precentor P. Joh. Gerner in his widely read beekeeping manual of 1881. Someone who loved heather honey, on the other hand, though not just any heather honey, was James Bond. He describes his perfect breakfast in *From Russia with Love*: strong coffee, a boiled egg, toast, Cooper's orange marmalade, Tiptree Little Scarlet Preserve and Norwegian heather honey from that home of tradition Fortnum & Mason. Norwegian! Karin would have loved to know that.

Norwegian chestnut honey would have been even better, of course.

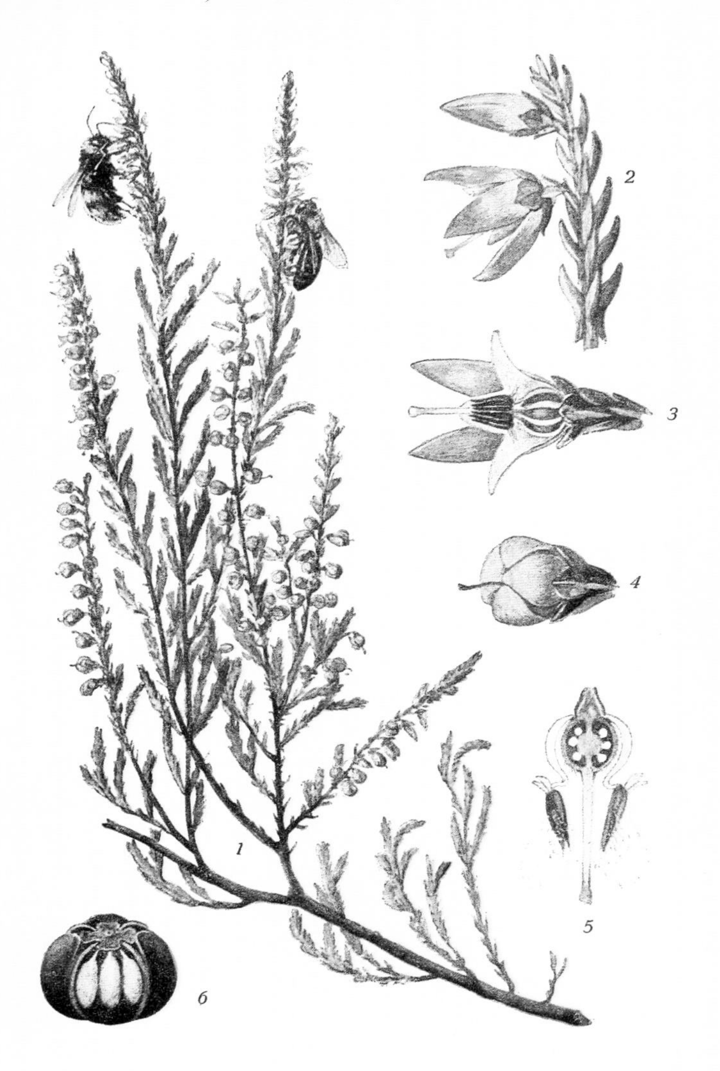

Heidekraut (Calluna vulgaris).

Honey from heather, Calluna vulgaris, *is loved by some and loathed by others.*

THE TASTE SHOULD DECIDE THE MATTER

A letter to the editor on the subject of heather honey was published in the March 1914 issue of the Swedish magazine *Bitidningen* (The Bee Magazine) along with a carefully considered reply.

" During the long nights of Christmas, as I was studying the literature on bees, I was struck by a sentence that stated that heather honey is the best sweetener in the world. I had previously heard only that this honey was somewhat inferior to pale honey in general. This inspired me to research the available material in order to discover the generally accepted view on the subject.

There does not appear to be any heather honey in America, as all the writers there are silent on the subject of this plant. In contrast, every English writer is full of praise for heather honey. One writes: 'The Scots – who by the way will have nothing to do with any honey other than heather honey – all take the view that while honey made from bell heather is delicious, that made from common heather is infinitely superior.' In another work by one of England's most highly regarded honey experts it is stated that 'Heather honey is one of the finest varieties of honey and that gathered from *Calluna vulgaris* is the very best.'

Those connoisseurs, the English, appear therefore to regard it as something quite exceptional. As honey of this kind is not available to those of us who live on the plains of Scania, it would be of the greatest interest to find out the generally held view in our country from those who have enjoyed this and other varieties of honey. "

J. Byman

The editor's response:

"Determining whether honey derived from a specific variety of plant is superior to another kind is unlikely to be without difficulty. A good many people attach value to the colour and allow the eye to evaluate the product. Using this form of appraisal, the pale lively gold colour is likely in most cases to carry the day. Many people also assess honey on the basis of its medicinal properties, and heather honey is considered superior in this regard. If, yet again, honey is being considered as a foodstuff and it is taste alone that determines its value, then the verdict is likely to be very varied. There are those who cannot eat heather honey, while others will only buy heather honey year after year, asking for it alone and not wanting any other kind if heather honey cannot be obtained."

Heather Bee, *etching by Laney Birkhead, a beekeeping artist from Yorkshire. The aim of her fascinating "swarm projects" is to increase awareness of the plight of the honeybee.*

The way honey tastes, smells and what colour and consistency it possesses are determined by the kinds of flower whose nectar the bees have gathered before producing it and storing it in wax combs. In order to produce varietal honey, pure honey from a single source – such as that of the lime, or linden tree blossom, which bees love – the nectar is harvested immediately after the flower has finished blooming.

AUGUST

A Honey Tasting and a Lecture on Honey Remembered

and

A Description of Today's Honey Fraud

S.S.B.F., THE BEEKEEPERS' ASSOCIATION of Southern Sweden, held its annual meeting at the Sjöhusgården café in Eslöv. Those of us who are members all brought along a jar of this year's honey whose taste, consistency, colour, purity and several other qualities were to be assessed by the association's three honey judges. Unfortunately only one of them turned up.

"The others must have swarmed off somewhere," said the association's chair Elof Nilsson, who had rattled over from Igelösa on his moped.

So the committee members had to come to the aid of the presiding judge, Malte Persson. Because it was a very hot day we were sitting in the shade of a large beech tree. Our treasurer, Inga Larsson, handed out spoons of honey that we proceeded to taste in a reflective manner.

Elof Nilsson, chairman of the S.S.B.F., is tasting Malte Persson's honey. Has it been strained enough? Until the end of the twentieth century it was generally believed, and not just within the S.S.B.F., that it was possible to remove any and all traces of pollen by straining the honey. But it turned out that this can only be achieved by using a technically advanced high-pressure filter. It is, however, no longer considered a bad thing if honey does contain pollen. Pollen is good for you, after all. Nowadays you can even buy honey with extra pollen added to it.

"This is an excellent honey," Knud Madsen says while other tasters look skeptical. Could it possibly be his own? You never know with Danes and that odd sense of humour they have.

"This jar doesn't get full marks," Malte said, "It's got too much pollen in."

The rest of us thought the honey had been properly strained. But Malte stood his ground.

"It's my own honey so I should know what's wrong with it."

You might think a grain or two of pollen shouldn't make any difference, especially as pollen is supposed to be good for you. I once interviewed Gösta Carlsson, known as the Pollen King, who became a millionaire thanks to his method of collecting large quantities of pollen straight from the field. According to him, he developed this technique based on the instructions he received from some extraterrestrials he came across in a forest clearing outside Ängelholm, where their U.F.O. had been forced into an emergency landing. Cernelle, his company, was now putting pollen into everything from nutritional capsules and remedies against ageing, to cosmetics, healing ointments and toothpaste. He was devoting the millions he earned to projects such as building an ice rink, sponsoring the hockey team Rögle B.K., and buying uncut diamonds in Amsterdam. He would then polish them in his basement workshop at home. What a man!

But it makes no difference however good for you pollen may be. A jar of honey may only contain pure honey according to the rules of the S.S.B.F., so Malte's jar could not be given full marks.

When the meeting's agenda was completed and once the sitting members of the committee had been re-elected, Erik Lassing, a teacher and one of the association's founding members, gave a lecture on skulduggery in the honey industry. As early as the sixteenth century Olaus Magnus was warning against imported honey that had been adulterated by cunning foreign merchants, he said, and there have been dirty tricks of this kind ever since.

ARTIFICIAL HONEY: THE INGREDIENTS – POTATO – SULPHURIC ACID – COAL TAR

Fake honey and honey substitutes are by no means rare. The crude fakes of the past that were made up of flour, starch, dextrin, cane sugar, molasses etc. are very unlikely to appear nowadays given the means currently available for detecting them. The use of a mixture of glucose and artificially produced invert sugar is, however, more difficult to prove. (Nordisk familjebok, *Swedish family encyclopedia)*

Sugar was rationed during the Second World War, but if you had bees you could get an extra allowance. The "sugar-beekeepers", as they were known, acquired empty beehives that they then registered as belonging to them in order to qualify for the extra ration which they then sold on or consumed themselves.

The sugar-beekeepers of more recent times have a different approach. They feed their bees with sugar all year round and then sell their adulterated honey as the genuine product. Though there are no longer any of them left in Sweden, or so Erik Lassing believes. He went on to tell us that so-called artificial honey is a different matter. It is produced entirely without any contribution by bees and can be made up of such things as invert sugar, syrup and artificial colourings and flavourings.

Everyone agreed this was a most interesting lecture. Afterwards it was time for coffee, waffles and conversations about the summer. A miserable one, or so almost everyone thought. There had been one bee swarm after another and there had not been much honey.

The best honey is your own

Along with olive oil and mozzarella, honey is one of the most adulterated foodstuffs in the world. A leader in this field is China, which is also the leading exporter of honey, frequently in collaboration with shady companies in other countries. One way of adulterating honey is to harvest it while its water content is still high, before it has had a chance to evaporate – a process often accelerated by the fanning of the bees' wings – and be capped with wax. Instead it is dried in machines and ultra-filtered to remove the pollen which would reveal where it came from. Then sugar is added, or syrup, water, colourings and flavourings plus starch, along with pollen from a different country which is then labelled as the country of origin. The ingredients listed in one of the less appetising concoctions are soya sauce, dead bees, beeswax and aluminium sulphate. Rotten rice may also be added.

Some Swedish beekeepers also adulterate their honey. It might say heather honey on the label even though the jar also contains honey from other flowers. A small profit can be made in this way, because heather honey is more expensive than other kinds. Not nice, though fairly innocent when viewed in a global context. Much less nice – and more common and certainly more profitable – is to market honey as manuka honey when it isn't. Genuine manuka honey is produced from the tree *Leptospermum scoparium* which only grows wild in New Zealand and the south of Australia. It is said to be exceptionally healthy because of its antibacterial properties and is one of the world's most expensive honeys. As a result, manuka honey is sold in far greater quantities than are actually being produced.

The best way of avoiding adulterated honey is to have bees of your own. The next best thing is to buy direct from a reliable beekeeper.

Winnie-the-Pooh, the best-known honey lover in literature. Drawing by Ernest H. Shepard.

SEPTEMBER

A Tricky Question Recalled

and

Repeated Attempts to Understand Rudolf Steiner's Thoughts on Bees

THERE'S SOMETHING VERY SPECIAL about your own honey for someone who has just become a beekeeper. I am contentedly counting my jars and feeling like Winnie-the-Pooh. Although I don't really like giving any of mine away, my sister got two pots. She loves honey and thinks it is wonderful that I have my own honey producers. She asked me recently how I go about taking honey from the bees, "harvesting", as we bee people call it, and how the bees manage without it.

"First I blow smoke from burning sacking onto the part of the hive where the honey is," I explained, "and the bees think that there's a fire, which means the priority is to save the honey at any cost. Once they've gathered on the honey frames, lifting them out and brushing the bees off is easy enough."

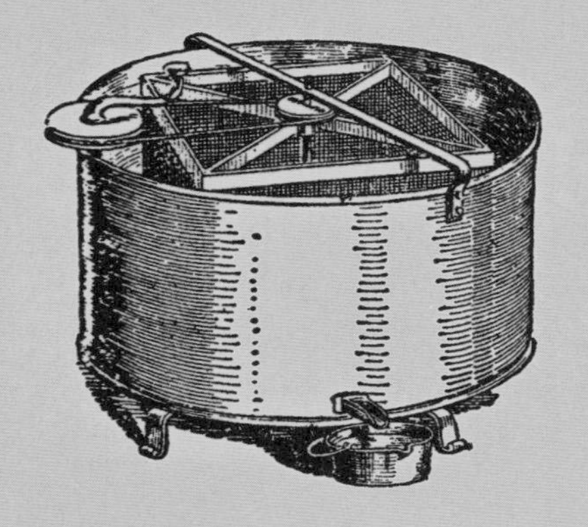

THE PIECE OF EQUIPMENT THAT CHANGED BEEKEEPING FOR EVER

The honey extractor was invented in 1865 by the Austrian army office Franz Hruschka, who had by then become a beekeeper in Italy. It revolutionised beekeeping which could henceforth be done on a much larger scale. The illustration shows Hruschka's own model. Today there are electric extractors for amateur beekeepers along with more technically advanced models for the professional bee farmer.

Before the advent of the extractor the combs were wrapped in linen cloths which were then squeezed until the honey ran out. This painting by C. G. Bernhardsson, who portrayed the lives of the common people in Bohuslän, shows the process in action. The wax combs could also be broken apart and the honey allowed to trickle out on its own. This method meant there was less wax in the honey, but it was more time consuming.

"Isn't that a bit mean, though?" my sister asked. I assured her it wasn't in the slightest. Even the ancient Romans did it that way.

I went on to describe how the honey frames are placed on a holder in an extractor, a drum or barrel turned by a crank that makes the frames rotate. The honey is centrifuged and flows out through a tube with a tap. Then as autumn approaches every colony gets approximately fifteen kilos of sugar dissolved in water that the bees can live on until spring arrives.

My sister leans towards anthroposophy and sent her children to a Waldorf school, and when she heard this she got upset. How can you deprive bees of a natural product containing minerals, enzymes, vitamins and many other nutrients and then force them to make do with sugar water? If that was the way bees got treated, she wasn't going to eat any more honey. Well, my jars of course, but then no more!

I didn't know what to reply to that. I had never thought about it, I had just done as I had been taught by John. All the beekeepers I had met gave their bees sugar solution. But the question stayed with me and wouldn't let me be, so I eventually bought *Über das Wesen der Bienen* (On the Nature of Bees) by Rudolf Steiner. It might have some important insights to offer that my bee friends did not know about.

RUDOLF STEINER (1861–1925)

The father of anthroposophy, biodynamic agriculture and Waldorf education. His ideas about bees were unlike those of any writer before him, to put it mildly.

Having now read it, or tried to read it, I have to admit that I don't understand much of what Steiner refers to as cosmic energies and their influence on life on earth, including bee colonies. The fact that the book was in German did not make things any easier, even though I can usually read the language fairly well. But I did at least understand that if bees were allowed to retain the greater part of their honey, they would be protected against the stresses and strains, as well as the diseases, of winter. Even though my sister had not read the book she had managed to frame the issue in authentically Steinerian terms.

On the whole it seems that bees should largely be left to manage themselves. They should be allowed to create their combs according to their own designs and not on prefabricated partitions. And swarming should not be prevented by removing queen cells or clipping the wings of the queens.

As someone who likes my bees best when I can watch them flying in and out of the hives undisturbed, that should really suit me down to the ground. You can always buy honey, after all.

Did Steiner ever keep bees of his own?

Because many of today's alternative beekeepers refer to Steiner, I have now read his book in an English translation in order, or so I hoped, to get more out of it. But even though the book was easier to read in purely linguistic terms, it remains one of the least comprehensible books I have ever come across.

So, if I were obliged to try and summarise it: everything that takes place on earth from the largest to the smallest occurs under the influence of the cosmos. As far as bees are concerned this means that the period between newly laid egg and full-grown bee is of key importance for their role in the colony.

It takes a worker bee twenty-one days, approximately the same length of time that the sun takes to turn on its own axis,

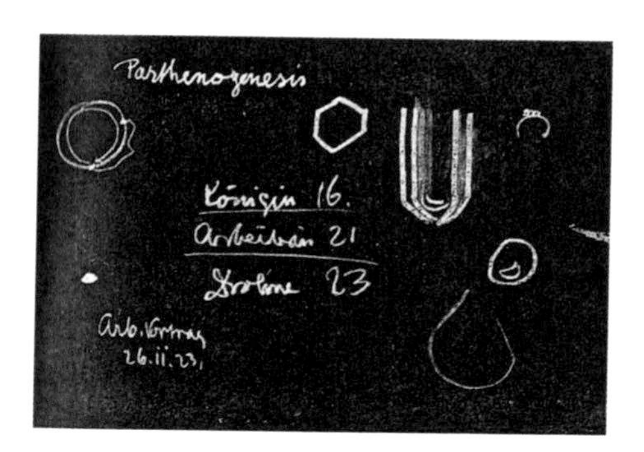

On the left, the development of the bee from egg to full-grown bee, chalk drawing by Rudolf Steiner on the blackboard of the Goetheanum. On the right, a 1905 portrait of Rudolf Steiner.

to emerge from the brood cell. This means – according to Steiner – that all the cosmic influence it has received came from the sun, and the worker bee is therefore a creature of the sun. The queen, which only needs sixteen days before she emerges, is also a child of the sun. That she is able, unlike the worker bee, to lay eggs is because she is closer to the larval stage. The accepted explanation: that a "normal" egg matures into a queen because the larva is fed with a special blend of nutrients known as royal jelly is not something Steiner mentions.

The males, the drones, do not emerge from their brood cells until twenty-three days have elapsed. This means they have time to enter a new cycle, one that follows that of the sun, and are therefore influenced by the power of the earth. This makes them earthly beings and capable of reproduction. The female capacity to produce eggs comes from the sun whereas the male capacity to fertilise comes from the earth . . . if I have understood the matter correctly.

It is not just the cosmic forces but also the hexagonal shape of the cells that plays a vital role in the process by which bees become bees. "The larva receives the forces of the *form* later it feels in its body that it was once in this hexagonally formed cell, in its youth when it was quite soft. [. . .] There lie the forces through which the bee afterwards works, for what the bee *makes* externally lies in its environment."

Does that sound odd? If you want to understand everything in Steiner's book without immersing yourself in the conceptual universe of anthroposophy, you might as well put the book down. The alternative is to accept that most of it is more or less peculiar though fascinating and thought-provoking, and go on reading.

The fact that honey is healthy can hardly be a surprise to anyone. According to Steiner this is not because of the nutrients it contains, however. He rejects the modern scientific method of breaking things down into their component parts

Roy and Bettie Brewster outside Norian House in Plymouth, New Zealand. Like Steiner, Roy Brewster believed that the hexagon was a very special polygon. For him it was a gift from God and the best shape to live in. There is nothing rectangular in nature. Parallel lines and right angles create a world of wickedness and lies. In 1954 he began building Norian House in which anything that could be made hexagonal was. The house became a popular tourist attraction but when his wife died he tore it down after receiving a message from God.

and analysing them to find explanations of their effects. On the contrary, the health-giving qualities of honey derive from the fact that it contains the power of the bee to create hexagons, which they received from the cell they developed in, and this power or force is one human beings need as well. Eating honey helps the body to maintain its shape. He recommends that engaged couples eat honey as preparation for producing children because honey helps the child-to-be to build its skeleton. Older people also benefit from honey, although it should not be eaten in excessive quantities because that can make the skeleton fragile.

The force of ego-organisation is another important concept. It can also be found in bee venom and the fact that people who suffer from gout can be made well by bee stings is because their ego-organisation is poor. The bee sting gives it a welcome boost.

But what does all this mean in practical terms for the beekeeper? Steiner does not have much to say about this. Did he have any experience at all of keeping bees? Probably not, and besides, how would he have found the time with all his writing and other activities. Pedagogy, reincarnation, astral bodies, aetheric bodies, biodynamic agriculture, the development of different races, eurythmics and so on. He does refer frequently, however, to the farmers in the region of the Austro-Hungarian empire, modern day Croatia, where he grew up. They did not keep bees in order to sell the honey and make money, but for their domestic needs. It was part of life in the country. What honey they did not need for themselves, they gave away to other families such as the Steiners. They never went without honey in his impoverished childhood home.

The few pieces of practical advice he gives are based on traditional methods that had by then been subordinated to the principles of modern beekeeping. Swarming should not be prevented and bees should be allowed to shape their own combs without any prefabricated partitions that determine the size of the cells. During harvesting, enough honey should be left for the bees to overwinter on. So no sugar solution, unless one is forced into it by a long winter. In which case one should add camomile tea and a bit of salt.

In Steiner's view, the most dangerous aspect of modern beekeeping is that the queen bee gets replaced before she has become old enough for her egg-laying to have diminished. Worker bees can never develop the same close links to rulers from outside as they share with those they have nursed themselves. This custom may prove disastrous, he predicts. If it continues then it will mean the death blow for beekeeping within a hundred years – by 2023 at the latest.*

The rules of the biodynamic Demeter association for beekeeping and bee products do not mention ego-organisation, sun or earth creatures or the cosmos, but stick to purely practical matters. For example, although the sugar that is used in any winter feed should be Demeter-certified, at least 10 per cent by weight of the winter feed must consist of honey. The bees must be allowed to build their own combs in the brood chamber without any prefabricated dividers. Replacing queens, preventing swarms and queen excluders, which stop the queen from laying eggs in the honey super, are not permitted either. New colonies and queens are to be produced by swarming, i.e. when the bees themselves want to propagate as a colony. "The key thing is how the bees are looked after and to what extent they are permitted to pursue their natural behaviour."

*Today, non-anthroposophical beekeepers and researchers also caution against replacing queens. It leads to reduced genetic diversity and thus to diminished immune resistance on the part of the bees. It also allows diseases and parasites to be spread by the international trade in queen bees.

HONEY ON YOUR FACE MAKES THINKING LIVELY
The German artist Joseph Bueys was greatly influenced by Steiner's ideas. The English edition of Steiner's book on bees ends with a special chapter on Beuys' anthroposophically inspired artistic alchemy. One of his most celebrated performance pieces, Wie man dem toten Hasen die Bilder erklärt *("How to Explain Pictures to a Dead Hare"), took place in 1965 at the legendary Galerie Schmela in Düsseldorf. He walked around for three hours with his face covered in honey and gold leaf and with an iron slab tied to one of his boots and a dead hare in his arms while he mumbled explanations to the animal of the drawings that lined the walls. The audience were not allowed in, but they were able to observe the event through the windows of the gallery. He considered the ability of bees to produce honey as corresponding to the human ability to produce ideas. "In putting honey on my head I am clearly doing something that has to do with thinking. [. . .] In this way the deathlike character of thinking becomes lifelike again."*

SAINT AMBROSE, THE PATRON SAINT OF BEEKEEPERS (339–397 C.E.)

It is said that when he lay in his cradle a bee swarm settled on his mouth and as it flew away it left behind a drop of honey. This was seen as a sign that he would be a distinguished speaker, which he was. He was known therefore as the "honey-tongued". The Catholic Church has traditionally had a special relationship with bees. For long ages the monasteries were both centres of beekeeping and repositories of knowledge about bees, and it was completely natural for the monks who constructed Buckfast Abbey, which Brother Adam would enter, to acquire bees.

OCTOBER

Brother Adam Remembered

and

His Posthumous Reputation and the Scandal He Managed to Avoid

IT IS 1996, AND THE BENEDICTINE MONK Brother Adam is dead at the age of ninety-eight. He was world famous for breeding the Buckfast bee, which is said to bring together all the good qualities a bee can possess. It is more resistant to parasites and diseases than other bees. It is enormously productive but not particularly given to swarming – disinclined to swarm as beekeepers say – and remarkably docile. Docile is the term used for bees that will sting only out of necessity, even if they were not actually bred by a cleric.

I have been reading the obituaries in the English newspapers. They refer to the many prestigious honours he was awarded, including the Order of the British Empire and the Bundesverdienstkreuz (Order of Merit of the Republic of Germany). He was also made an honorary doctor at the Swedish University of Agricultural Sciences in Uppsala, an

Brother Adam in one of his apiaries meant for honey production. The breeding of Buckfast queens takes place on Dartmoor itself.

honour he was particularly proud of because it was awarded by an academic body. "He was unsurpassed as a breeder of bees. He talked to them, he stroked them. He brought to the hives a calmness that, according to those who saw him at work, the sensitive bees responded to," *The Economist* writes.

And to think that I have met this giant among men. After so many years I ask myself how I could ever have dared to write and ask if I could visit him in the monastery at Buckfast Abbey. I knew nothing about breeding bees and, even though I had bees of my own, my scientific knowledge about the species *Apis mellifera* was rudimentary.

As a journalist you cannot, however, allow yourself to be intimidated by a sense of your own limitations. That will get you nowhere. Besides, my situation at the time was a peculiar one, which may have made me cockier than usual. For several years I had been employed at *Femina*, which had by then made a name for itself as a radical women's magazine. But now the owners, the Danish Aller family, had had enough. "We're taking our red stockings off," they announced in full-page ads in the major newspapers. "We want to be your break from the debate!" The offending editorial team had disappeared, but a new one had not yet been appointed. This meant that new ideas could be tried out that would have been blocked in other circumstances by any responsible editor. The main objective for the provisional management was to fill the pages of the next issue, and because *Femina* was a weekly magazine at the time rather than the monthly it is now, a constant stream of ideas was vital. And most of them were well received, provided they had nothing to do with women's liberation.

I had made up my mind to resign, but I wanted to go to England first – paid for by the magazine, of course – to write about 1.) a walking trail along the coast of Cornwall I had seen all these tempting pictures of, and 2.) Brother Adam, who I had read about in *Bitidningen* (The Bee Magazine) and had thought it would be interesting to do a profile of him. The mere fact that he was eleven years old when he was sent from a

When Brother Adam arrived at Buckfast, work on the new abbey was still under way. But because he was a sickly boy, he did not have to undertake the heavy labour involved in the construction. He was allowed to help with the abbey's bees instead.

little village in southern Germany to a monastery in England: What can that have felt like?

I got the go-ahead for both of these rather odd projects. Those were the days! I wrote to Brother Adam and he replied that I was welcome to visit him. At Heathrow I was met by the formidable Lena Svanberg, the then London correspondent of *Veckans Affärer* (a Swedish business weekly), who had heard about my walking plans and let me know that she planned to join me. As it happened I had been looking forward to a meditative week listening to the waves and the wind and contemplating what I was going to do with my life after *Femina*. But when Lena had made up her mind to do something, she could no more be deterred from doing it than the sun from rising, and it is true that she was far from boring company, quite the contrary. Having her car available was a practical advantage too. I would just have to do the meditating afterwards.

Buckfast Abbey is in Devon, close to the main road to St Ives in Cornwall where our walking tour was to begin. So hardly a diversion at all. My suggestion that Lena should join in on the interview and write about the triumphal progression of the Buckfast bee in *Veckans Affärer* was promptly rejected. The only creatures that interested her were horses – she could bet on those.

"You can keep that monk to yourself, but I'll pick you up at four on the dot," she yelled as she dropped me off at the monastery. She planned on making the local pub, The Abbey Inn, her headquarters and reading old newspapers, a favourite occupation of hers. Wherever she went she took a plastic bag full of unread newspapers with her.

The monastery turned out to consist of an enormous church and a collection of smaller buildings, all of it hideously nineteenth century in style. But Brother Adam, even though he too was from the nineteenth century – born in 1898 – was a handsome, quiet and kindly man. I realised very quickly that an intimate portrait was out of the question. It was not the fact that he was a monk. I had met monks before and some were as talkative and open about themselves as the writers and actors I was used to interviewing. But Brother Adam was entirely focused on bees and appeared to believe I was too. There was a gravity about him that inspired respect, no doubt amplified by his German accent, and that was not conducive to personal questions.

He said he was sorry he couldn't show me his queen farms up on Dartmoor because all the monastery's cars were being used. Lucky for me. Because then my obvious lack of genetic knowledge would have been embarrassing. Instead he took me with him to one of the apiaries that were set up for the production of honey. The neatly painted hives stood in groups of four, each group turned to a point of the compass. Not in lines which is otherwise the norm.

"That's to help the bees find their way and stop them entering the wrong hive," Brother Adam said. "If they do that they'll end up getting stung to death, after all."

What loving care! Why don't all beekeepers set up their hives in this way?

"Because it's more practical and takes up less space to arrange them in rows," he replied. Just as I was about to photograph him, a bee came flying along and settled on his shoulder, as though to demonstrate its affection.

Then we visited the spotlessly clean premises where the honey and the wax are prepared. The equipment included a gigantic extractor Brother Adam designed himself for heather honey, which can be hard to extract. The area was not currently in use because it was too early in the year for any honey to be harvested.

Our tour ended in the monastery's very own little bar. Brother Adam got out a bottle labelled *Buckfast Tonic Wine*

Brother Adam pouring the monastery's tonic wine which has gained a cult following among some groups of young people in Scotland.

and poured two glasses. We raised our glasses in a toast. The wine tasted a bit like vermouth, sweet but not sickly.

"To your health," he said, and added that this really was a healthy wine. The recipe arrived with the French Benedictine monks who came to Buckfast in the nineteenth century, after they were driven into exile by the anti-Catholic Third Republic. They reconstructed the monastery which had been torn down during the Reformation, and to earn their daily bread they kept bees and prepared medicines, salves and this wine. Although now, he added, most of the monks, including himself, came from Germany.

At last something to get my teeth into! I could now ask him directly about his background. Where in Germany did he come from? And what did it feel like to be sent to a monastery in England as a little boy, and why did he stay here and how had that affected his faith? Was it compatible with genetic research?

But that wouldn't work. It simply was not possible to come out with those kinds of questions. Instead I asked what it felt like when he decided to breed a new kind of bee, a question he must have been asked numerous times before.

He topped up our glasses. "It began with a catastrophe," he said, without seeming the least bit bored or impatient.

During the First World War the monastery's apiary was affected by the devastating tracheal mite that had already exterminated most of the native English bees, he told me. The few that survived turned out to be either yellow Italian bees, *Apis melliferia ligustica*, or hybrids of the Italian and the native dark bee, *Apis mellifera mellifera*. It was at that point that he crossed new Italian queens with drones from the resistant colonies. That turned out to be the right track. One of the new Anglo-Italian colonies – only one – possessed all the good qualities of both the Italian and the English bees but none of the bad ones. Its queen became the mother of the Buckfast bee, its Eve. After the Second World War he had continued to add good genes to this hybrid bee that he had discovered on the lengthy journeys he had made.

"From Sweden as well?" I asked.

No. Although he had of course been sent native Swedish queens, the offspring were dreadfully inclined both to sting and to swarm, and it made no difference that the breed was highly productive.

It was approaching four o'clock. I thanked him for an extremely instructive visit, we raised our glasses in a final toast and then he accompanied me to the gates.

"Remember you have to listen to the bees," he said. "They follow their own desires and not ours."

Lena was waiting out on the road.

"You smell of booze," she said when I got in the car. "So what have the two of you been up to?"

"We drank tonic wine," I said.

We were almost in St Ives by the time she had finished laughing.

My article for *Femina* about Brother Adam didn't turn out particularly well, but the photograph of him got a whole page.

The second time I met Brother Adam was in the Frostavallen recreation area in southern Sweden a few years later. The Queen Breeders' Club of Kristianstad had invited him to give a talk and close to a hundred people turned up. I had Annicka Lundquist for company, my very first bee-acquaintance.

Brother Adam recognised me and greeted me warmly. I was rather proud of this and hoped the people around us had noticed. But was it really true, as the papers had said, that thieves had stolen his valuable breeding queens?

"Sadly, yes," he said, "you can't keep an eye on what people get up to up there on Dartmoor. But they only made off with two queens plus combs containing workers and drones – I hope they got badly stung."

And no, he hadn't been to Sweden before, but he had had a great deal of correspondence with Swedish beekeepers and knew that it was a good country for beekeeping with a varied wild flora and beekeepers who were better organised than in any other country.

The talk was about his fantastic travels in a rattling car and occasionally on foot or by mule to inaccessible places including sites in Spain, Turkey, Jordan, Egypt and Morocco, on the hunt for rare bee tribes with positive qualities. Genes from a Greek bee, *Apis mellifera cecropia*, had made the Buckfast bee even more docile and less inclined to swarm. Much of its hard-working nature derived from *Apis mellifera anatolica*, which he discovered high up on the Anatolian plateau. A Cypriot bee had contributed its ability to survive hard winters.

When it was time for questions one listener asked him where he wanted to travel to next.

"To Kenya and Tanzania, to look for *Apis mellifera monticola*," he replied, a bee that is very docile, unlike most African bees.

Next question: "Who pays for all these trips?" Answer: "The monastery does." I should think so too. The sale of Buckfast queens brings in a huge income from which the original beekeeper gets not a penny.

The last audience question was whether Brother Adam ate a lot of honey because he seemed so fit. Yes, he did. At least a tablespoonful a day, and preferably heather honey from Dartmoor.

The Consequences of the Vow of Obedience

Unfortunately there is also a sad side to the story of Brother Adam that would affect him in what were his final years. In 1990, the devastating Varroa mite was discovered in the monastery's apiaries. Naturally he wanted to produce a Varroa-resistant bee and for that he needed an assistant. But the monastery's newly installed abbot, David Charlesworth, refused his request. The apiaries existed for the production of honey and not for genetic research and, besides, the bees were a distraction for Brother Adam from his vocation of serving God.

A milder variety of Buckfast humour – a pastiche of the painting Babie lato *("Indian Summer") by Józef Chelmonski. Far less polite examples can be found such as a poster showing Jesus with a bottle in his hand and the legend: "It's my fucking birthday!"*

Voices all over the world were raised in protest, but Charlesworth made clear his conviction that Brother Adam should see himself as a monk first and foremost, and only secondarily as a beekeeper.

Brother Adam felt obliged to accept the decree because of his vow of obedience. He died four years later, by some accounts a bitter and disappointed man. But at least he managed to avoid all the uproar and commotion of recent years that have surrounded that same Buckfast Tonic Wine he once served me at the monastery.

It turns out that much of the violent crime in Scotland is committed by young people who have been drinking "Buckie", as the wine is called there, even on occasion using the empty bottle as a weapon. "The Buckie made me do it!" has become a classic defence argument. Consumption is particularly high in the so-called "Buckfast Triangle" between Airdrie, Coatbridge and Cumbernauld to the east of Glasgow, and so is youth crime. It is the combination of 15% alcohol with caffeine in quantities to rival several cans of Coca-cola that has made the wine into a cult drink among the socially excluded. There are T-shirts and baseball caps with images of Buckie as well as Buckie songs with filthy lyrics. Scottish politicians have repeatedly demanded that the wine be banned, but Father Charlesworth and the distributors Chandler & Chandler have argued in response that in that case the Scots would have to prohibit their own whisky and that sales of Buckfast wine account for less than 1 per cent of total alcohol sales in Scotland.

Buckfast wine is not for sale in Sweden.

The Buckfast Bee

The Buckfast bee, sometimes referred to as "the beekeeper's bee", is now bred widely across Europe, with the pedigree being regulated by the German-dominated *Gemeinschaft der europäischen Buckfastimker* (Federation of European Buckfast Breeders) operating through twenty-six countries. Sweden is the only country where the Buckfast bee is predominant, and the *Föreningen Svensk Buckfastavel* (the Swedish Association of Buckfast Breeders) has its own breeding stations on the Swedish islands of Aspö, Hässlö, Ven and Vendelsö. Isolated mating stations were first established in the U.K. in 1971. The above bee is a Buckfast queen.

No matter what its subject, no historical exhibition in France can avoid mentioning the Revolution even if only in some corner of the show. This was also the case for an exhibition about bees, people, honey and wax held at the Musée des Arts et Traditions Populaires. In this image of Revolutionary liberty on display a beehive symbolises the national assembly in which the Church, the nobility and the Third Estate all collaborated. Not that the collaboration would continue for long.

NOVEMBER

A Trip to Paris Remembered

and

A Report from the Village of Älghult

I HAVE JUST GOT HOME from Paris and am rushing out into the garden to see my hives. The landing boards and entrances are empty. The bees are bound to have settled inside in their winter clusters and must not be disturbed. I gave them their food for winter before I left.

What was I doing in Paris? The trip included a visit to a fascinating exhibition at one of the ethnographic museums there: "L'abeille, l'homme, le miel et la cire" (The bee, humans, honey and wax). So it wasn't just about bees, their behaviours and their products, but also about the nature of the relationship between human beings – mainly the French – and bees through the ages. The catalogue was very scholarly and informative.

There were old hives made of clay, limestone, hollow wooden logs and straw on display as well as various kinds of protective clothing. The most elegant was a linen tunic with a striped visor made of horsehair. According to a quotation from one eighteenth-century French writer, however, no protection is needed if you wash your hands in hot urine and blow smoke from burning white linen into your face.

There were also many examples of the various uses to which honey and other bee products have been put. Before sugar became a staple in the nineteenth century, honey was the only real sweetener available and it was used extensively in cooking. Above all, though, it was a key ingredient in mead – *hydromel* in French – the very first alcoholic drink made by man, which has been regarded as a drink of the gods in most cultures. The Egyptians, the Greeks and the Romans drank mead, and so did the Vikings, of course.

The oldest known and simplest recipe for mead was written by Aristoteles in 350 B.C.E. Water and honey are allowed to ferment, and that's it.*

The ancient Roman Columella's recipe is far more complicated. Rainwater that has been allowed to stand in the sun for several years is poured into another vessel without the lees being transferred as well. A sextarius (half a liter approx.) of this water is mixed with a pound (roughly quarter of a kilo) of the best honey. The mixture is poured into a stone bottle which is sealed and left to stand in the sun for forty days while the dog star is rising, and is then moved to an attic to which the heat rises from the hearth. I can't help wondering what this drink tasted like and whether it was worth the effort.

Chouchen, a special variety of mead, is still made in Brittany from buckwheat honey that is fermented together with apple juice. This drink was said to have been introduced by the

*Mead has become a fashionable drink once more thanks to Harry Potter and Madam Rosmerta's mead and even more to *Game of Thrones*, which is said to have tripled its sales worldwide.

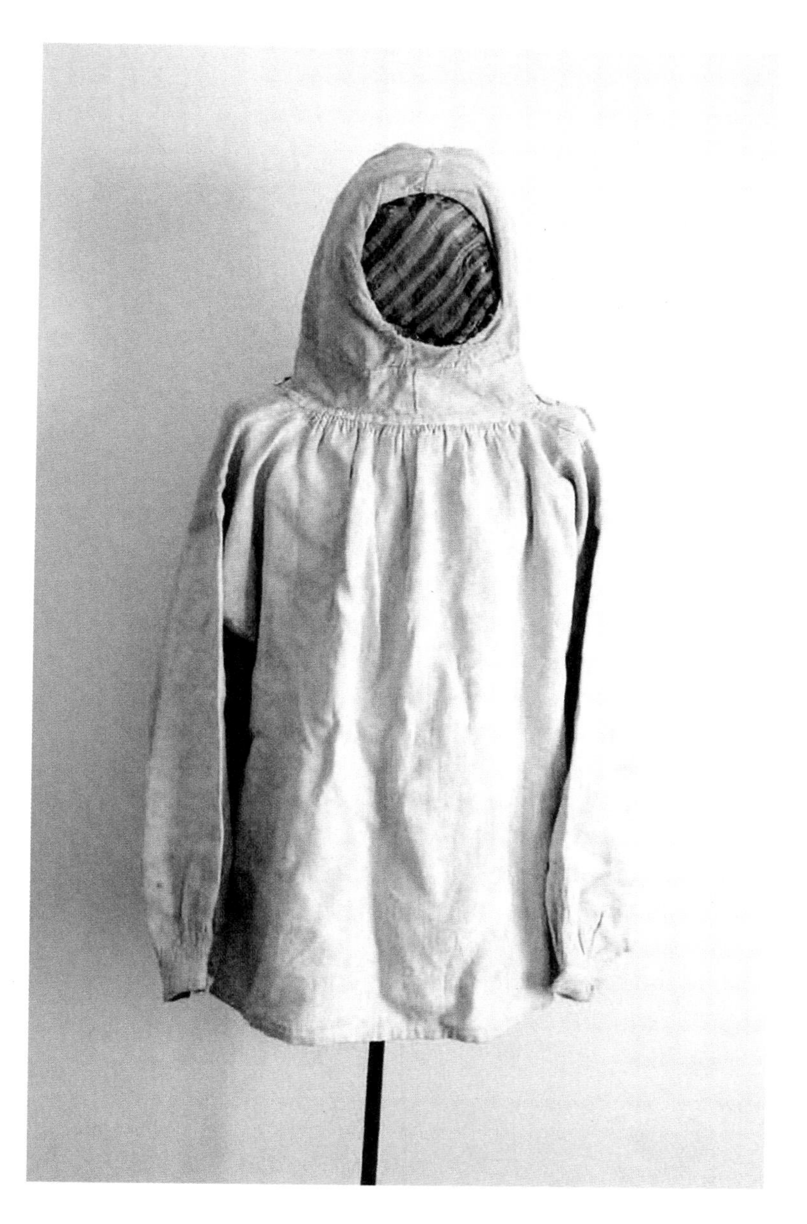

Reconstruction of a medieval jacket for beekeepers, with a protective visor made of horsehair. Very like the clothing worn by the figures in the Bruegel drawing on p.95

Traditional French treatment for rheumatic pains. Nowadays the venom is usually injected.

Celts when they were driven out of what is now England by the Anglo-Saxons. Unfortunately no samples were on offer.

Medicine was the next theme and it proved to be extremely interesting. Since the earliest times honey has been something of a universal remedy. Hippocrates, the father of Western medicine (460–370 B.C.E.) declared that the use of honey against all kinds of diseases was so widespread that some people had even begun to dislike the taste. Honey has been prescribed for ulcers, heart problems and lung disease. Applied externally while still fresh and runny it has been used, and is still used in many countries, to heal wounds.

It is a pity the curators were not familiar with the *Kalevala*, the Finnish national epos, in which the mother of the rogue Lemminkäinen pieces his limbs together after he was hacked apart while trying to kill the Swan of Tuola. Just how is she going to get them to join up, though? With honey, of course.

> "Tiny bee, thou honey-birdling,
> Lord of all the forest flowers,
> Fly away and gather honey."

It would have suited the exhibition perfectly and would no doubt have sounded wonderful in French as well.

It is not just honey that has been used to treat ailments and diseases – rheumatic conditions by and large – but bee venom as well. An old photograph shows a simple treatment method: a man whose upper body is bare is leaning forward while another man places bees on his pale back.

I remember a woman I once met in Landskrona, in Sweden. She frequently suffered from aches and pains as a child, but was cured by having a couple of handfuls of bees poured into her knickers. Although the pain never recurred, she couldn't eat honey ever again. I was also told about a man who would visit a beekeeper once a year to be stung, to prevent him developing rheumatism.

Nowadays one can visit an *apithérapiste* and have the venom injected. In France at least.

Catalogue to the Paris exhibition: "L'abeille, l'homme, le miel et la cire" (Bees, people, honey and wax).

Mummy portrait in encaustic paint, beeswax and pigment, c. 150 B.C.E., when Egypt was a Roman province.

The next section dealt with wax. This was once at least as prized as honey. Greeks and Romans wrote with metal styluses on wooden tablets that were covered in a thin layer of wax. In encaustic painting the paint is made by heating beeswax and mixing it with pigment. The Catholic Church needed endless quantities of wax for the candles that were essential for many of the liturgical ceremonies, and most monasteries had their own apiaries, not for the honey but for the wax. Death masks were made of beeswax and when anatomical figures were first created for teaching medicine they were made of it too.

The final section revolved around folklore and popular customs. Close emotional ties were believed to exist between bees and their keeper, and when something important happened within the family, the bees had to be informed. It was customary in Brittany to tie a strip of the red bridal dress on each hive when the daughter of the house got married. It was particularly important that the bees were told of the death of their owner before anyone else. Black cloth was tied to the hives and verses were read out telling the bees that though their master was gone they would still be well cared for. If not, they either became impossible or simply died.

There really should be exhibitions of this kind all over the world. There must be a large number of interesting objects at museums and other centres, not to mention the literature of the past on bees.

The museum run by Uppvidinge Beekeepers' Association in the Swedish village of Älghult is the only one of its kind in the Nordic countries. The figure in the lower image is engaged in drumming, which meant using your hands to drum on a standing hive so that the bees inside moved into an empty hive above.

Älghult may not be Paris, but it is very nice indeed

It's been quite a while since that exhibition took place in Paris. Even so, as far as I know, no comparable exhibition has been put on in Sweden in the intervening years. There is, however, a Museum of Beekeeping in Älghult in the Swedish region of Småland, the only one of its kind in the Nordic countries. It is far from as comprehensive and professional as the exhibition in Paris, but it has other admirable qualities and is the perfect destination for a day out.

It is run as a charity by the beekeepers' association in Uppvidinge. The house, which is an Astrid Lindgren dream, has been placed at the disposal of the association by the owner. The members have papered the walls, laid the floors, collected the objects on display, put up stands and written the captions themselves. The outhouse has been painted honey yellow and bees have been embroidered on the towels. Honey, pollen, propolis and decorative bee objects are on sale in the little shop. The cakes in the café are home-made. The personal touches are everywhere and it feels well thought through. And besides, you can study real bees as they crawl through plastic tubes on their way to and from a hive inside the museum. Not something you could do in Paris.

For bee museums around the world see pp. 212–13.

Honey extraction in 1952. The frames of honey are placed in net baskets that are then centrifuged until the honey flows out through a spigot.

DECEMBER

A Mysterious Honey Remembered

and

The Art of Tasting the Difference between One Honey and Another

Freshly extracted honey is runny and clear. But sooner or later, depending on what flowers the bees have collected the nectar from, it begins to crystallise and becomes cloudier and thicker. It should then be stirred daily until it develops an even, creamy consistency. That is how we like it in Sweden. We don't want the honey crunching between our teeth when we spread it on our bread. In many other countries they are not bothered by it being grainy and if it is used as a cooking ingredient and instead of sugar in tea then it doesn't really matter in any case.

There was something strange about this year's late summer honey though. Previously I had been able to tap it into jars after a few weeks, but this time it did not even begin to set until November. It was also darker than usual and the taste was different, fuller and with a spicy hint of bitterness. Normally it has a slight taste of peppermint because most of the nectar has come from the lime trees that flower along the road here in July. (You would think that honey tastes like the smell of the flowers the nectar has been gathered from, but that is not the case.) What had my bees brought home with them?

Lennart Kuylenstierna's honey was different this year as well – our bees have the same foraging area – and after we conferred he went to see Professor Åke Hansson who wrote *Bin och biodling* (Bees and Beekeeping), the bible of many Swedish beekeepers. Like us, he lives in Lund, and Lennart is personally acquainted with him because he used to be a shoe manufacturer. Now that he is retired he arranges for shoes to be hand-sewn for people with sensitive feet, including Åke Hansson. That was what led in turn to Lennart becoming interested in bees and eventually getting his own.

If my mentor in the bee world is John Larsson, then Lennart's is Åke Hansson. The views these two gentlemen hold on the technicalities of beekeeping – such as how many frames you should overwinter your colonies on – differ considerably, however, as do their views on life. Hansson represents the scientific approach, whereas Larsson that of popular or folk wisdom, a celebrated idea in the '70s and '80s. One writes books; the other regards book learning with suspicion.

Now and then, though, when I am stumped and don't want to bother John, I turn to Lennart and if he, too, feels unsure, he will ask Åke Hansson. This has led to me finding myself in some sticky situations when John has inspected my hives afterwards. Why have I done this and that? Am I aiming to kill my bees? It feels as though I have been having an affair on the side. My advice to anyone thinking about keeping bees: choose one guru and stick faithfully to him or her.

And now back to our mysterious honey. Lennart asked Åke Hansson to analyse it and the verdict was as follows:

1.) The runniness is because, besides the lime trees in our neighbourhood, there are many robinias, *Robinia pseudoacacia*, also known as false acacias. They flowered profusely this summer and our bees gathered a great deal of nectar from them. Pure robinia honey is light in colour and is one of the varieties that never crystallises.

2.) The taste and the dark colour are due to the fact that it scarcely rained in July. Once the bees had finished with the robinias they would normally have moved on to focus on the lime trees, but because of the drought the latter's flowers produced much less nectar than normal. There were a great many aphids, on the other hand, that flourished on the sugary leaf sap. Having passed through their digestive systems it came out as a sticky sweet secretion called honeydew, which our bees were delighted to dispose of. They then produced this very distinctive kind of honey which is marketed by beekeepers as honeydew honey. If you were to mention what it actually contains – Lennart refers to it as aphid-poo honey – people might be less keen to consume it. Which would be a pity, as the taste is lovely.

What should we call this year's special mix of robinia and aphid-poo honey? "Advent honey" is Lennart's suggestion and I cannot come up with anything better. A present for the connoisseurs in the family this coming Christmas.

There is a difference between one kind of honey and another

“Honey is a balsamic substance that possesses the quality of drying all unnecessary fluids while also expelling them; it resists rotting, promotes urine, loosens hard mucus, strengthens the stomach, banishes swellings, cures throat ailments and counters coughs and several other infirmities. ”

From *Afhandling om skånska biskötseln*
(Dissertation on Beekeeping in Scania), 1759

Just as there are wine tasters who can tell where a wine comes from, so are there honey adepts who can identify the variety of honey they are being offered, and not just which flowers the nectar has come from but also where those flowers have grown. *Le terroir*, as connoisseurs put it, though that usually refers to wine.

How do you become an expert on honey? Frenchman Michel Gonnet was the first person to apply a technique to honey that has become known as sensory analysis.* In the 1980s he developed a training course in honey-tasting for the International Federation of Beekeepers' Associations, also known as Apimondia, and in so doing paved the way for the current interest in the taste qualities of honey.

The only institution in the world to provide an advanced course in the sensory analysis of honey is the agricultural research institute C.R.E.A. in Bologna, from which more than three hundred honey tasters – *assaggiatori ufficiali di miele* – have graduated. One of the few non-Italians who can boast of this title is the American Marina Marchese. She calls herself a honey sommelier and has founded the American Honey Tasting Society. She is also the author of *The Honey Connoisseur*. She recommends that the best way of learning to evaluate honey is to acquire small samples from reliable beekeepers and to compare them. You make notes on the colour, aroma,

*Sensory analysis is a technique that involves experts assessing the qualities of foodstuffs using sight, smell, taste, sensation and even hearing, and then using scientific techniques to evaluate their findings.

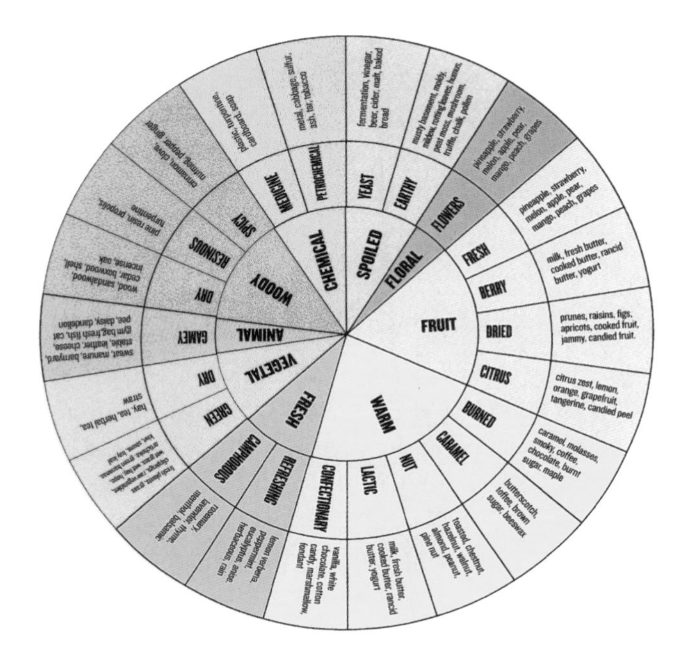

Template for honey tasting from The Honey Connoisseur *by Marina Marchese. You start by choosing from the more basic taste categories at the centre of the wheel – such as warm, fresh, animal, chemical, fruity – and then work your way outwards towards increasingly precise terms. A warm taste might be coffee, chocolate, toffee, roast chestnut or yoghurt. Or something completely different. An entire science of its own.*

consistency and taste. Best of all, of course, would be for several of you to get together and discuss the different varieties.

There were honey connoisseurs long before you could train to become one, of course. Pliny the Elder considered that the finest honey came either from the Hymettus mountains in Attica or from Hybla in Sicily, and the second best from the island of Calydna, which is called Kalymnos today. A more recent honey expert was John Milton, not to be confused with the seventeenth-century poet. He ran a shop on the Strand in London that sold honey and beekeeping equipment. He also

Apothecary jar of rosemary honey from Narbonne which was celebrated as far back as Roman times.

wrote *The Practical Beekeeper*, which was published in 1851 and contains the following passage in which he begins by referring to the honey "that obtains the highest price on the London market:

" This is the Menorca honey. It partakes of the flavour of orange flowers, is of a light gold colour, rather granulated, does not set very firm [. . .] It is imported in curiously shaped jars, having four short handles [. . .] The Narbonne honey owes its flavour to the many aromatic flowers that abound in that part of France, such as the wild thyme, rosemary, etc. [. . .] There is a white honey, very similar to this, but more granulated, that comes from Caen in Normandy. It is imported in casks and jars of various sizes [. . .] and in form not unlike a Roman pitcher. The valley of Chamouni is known to produce remarkably fine honey, which has an agreeable taste. No one flavour predominates. In this, I think, consists its adaptation to most palates. It is sent to us in very tastefully constructed vessels, or casks [. . .] West India honey comes chiefly from Jamaica. It is of excellent quality, rather granulated, and is much liked in England. [. . .] East India honey is imported in bottles. It is very fragrant, of rather a dark colour, and is not much esteemed in England, perhaps on account of its very powerful flavour [. . .] Seville honey possesses a delicate almond flavour, and never becomes firm; but remains thick and glutinous [. . .] The honey imported from the vicinity of the Pyrenees partakes of the flavour of the raspberry.

We receive very large importations of honey from Germany; but I am not aware that the honey of that country possesses any

TAL,

Om

LÅCKERHETER,

Både i ſig ſjelfva ſådana, och för ſå-dana anſedda genom Folkſlags bruk och inbillning,

Hållet för

KONGL. VETENSKAPS ACADEMIEN

Vid Præſidii nedläggande,

Den 3 Maj 1780,

Af

BENGT BERGIUS,

Banco-Commiſſarius, Ledamot af Vetenſkaps Sällſkap. i Trondhiem, Zelle, Lund, Götheborg, Nat. Curioſ. i Berlin, ſamt Patriot. Sällſk. i Stockholm.

Förra Delen.

STOCKHOLM,

Tryckt hos JOHAN GEORG LANGE, 1785.

Tal om läckerheter *(Dissertation on Delicious Foods) is one of the classics of Swedish gastronomic literature, in which Bengt Bergius describes the delicacies people have enjoyed since ancient times.*

remarkable quality. Honey is produced in great abundance in Russia, where there is, in fact, a separate market for it. One kind of Russian honey is very peculiar, and whiter than any other; it is of the flavour of penny-royal, which, with various other aromatic herbs, grows very luxuriantly in a wild state in the woods and wastes of this vast and partially cultivated country. When only slightly flavoured, this honey is agreeable; nevertheless, it is but seldom eaten by epicures. It is imported in casks turned from the stems of solid trees.

I have now mentioned many kinds of fine honey; but I think there is none that surpasses our fine English, particularly that which is obtained while the Dutch clover is in flower. This plant, so serviceable to bees [. . .] is found in great abundance nearly all over our island [. . .] We also receive honey of the finest quality from Northumberland and from the Highlands of Scotland.”

Several of the regional honey varieties that Milton praises are still recognised as particularly superior.

The Swedish historian Bengt Bergius (1723–84) was an expert on honey as well, though of a different kind than Milton. *Tal om läckerheter*, the mammoth lecture he gave to the Royal Swedish Academy of Sciences in 1780 dealt with practically everything on this earth that has ever been described as tasting good, from fried boa constrictor to elephant trunks. He had never been abroad, however, and can hardly have tasted any of them. On the other hand he had spent years researching what other writers thought of them.

He was able to confirm that in France the most superior honey was deemed to come from Narbonne. It was completely white, thick, granular and sweet, with an aromatic scent of rosemary. One of the connoisseurs Bergius had read asserted that Narbonne honey was the best in all of Europe, while another thought that the rosemary honey from Menorca was just as good, if not better.

He continues to provide descriptions of honey in this vein from Pliny onwards. Among his own contemporaries, one expert favoured Alpine honey from Appenzell in Switzerland, while another wrote warmly of the Spanish rosemary honey from Alcarria. One French botanist considered that the honey from Podor, in Senegal, "far exceeded in taste the best honey from the South of France".

Bergius, however, says not one word about Swedish honey. Unlike Milton he does not appear to have thought our own domestic honey was worth bothering with:

> “I take it as read that of the European honey varieties the most superior are those to be found in southern Europe, but the writers fail to agree on which varieties should be included under this label, and here in our country nothing can be done to conclude this debate as commerce provides us with only a few varieties from abroad of which the most refined comes from Provence.”

Was Bergius right? We have wonderful opportunities to sample different kinds of honey now that we travel more than ever. You really should have a go. It is always exciting to taste the local honey, whether you are in the Greek archipelago, Australia or southern Sweden. You can also find ultralocal varieties of honey in a number of places, thanks to the growth in popularity of beekeeping in towns. In Copenhagen the Bybi (town bee association) sells honey from various parts of the city. The honey from posh Frederiksberg is quite different from that produced in hip Vesterbro.

Most of the museums listed on pp. 212–13 sell local honey varieties. There are also many shops that specialise in honey and other bee products. You can find them on pp. 213–14.

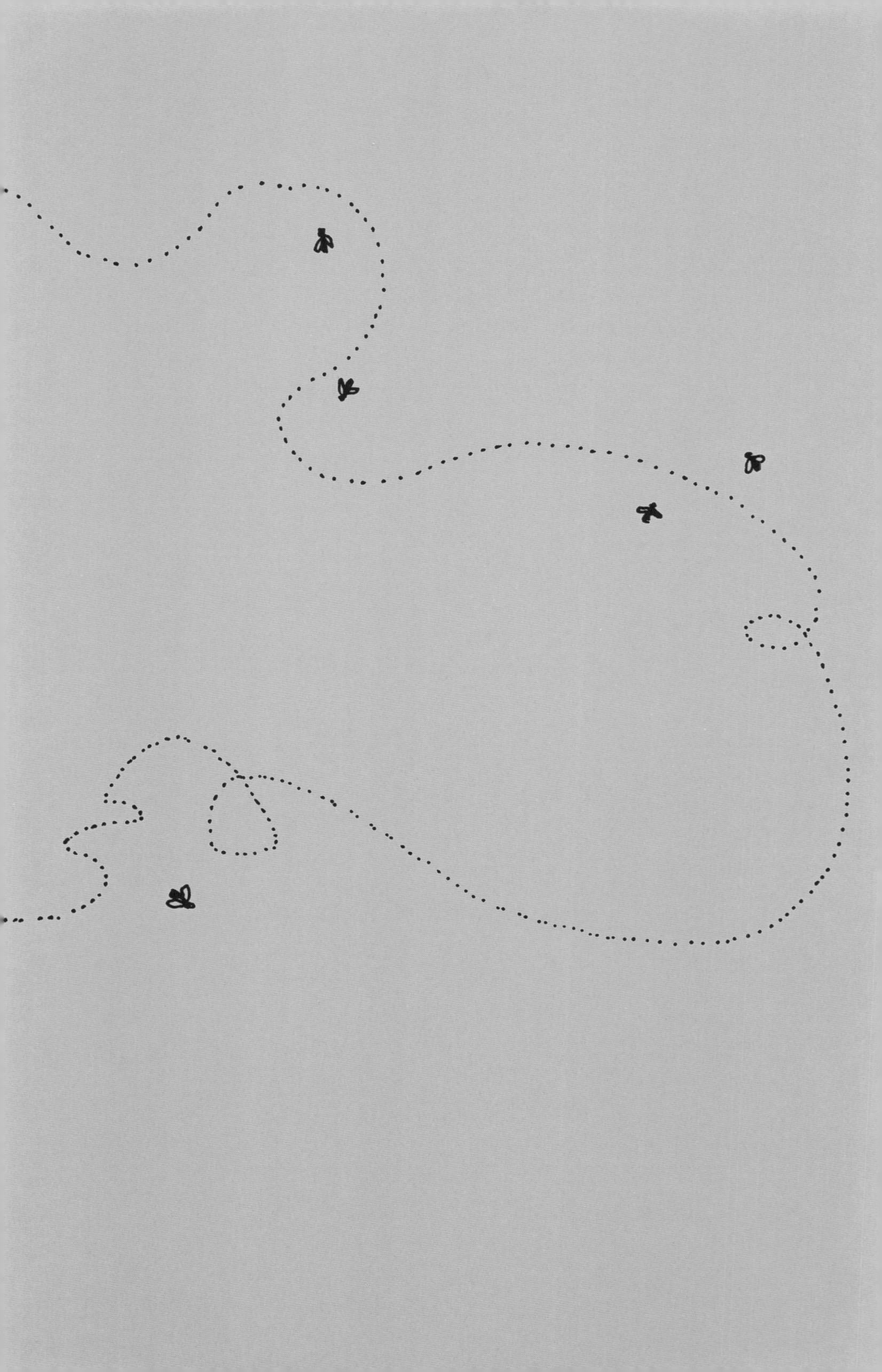

PART TWO

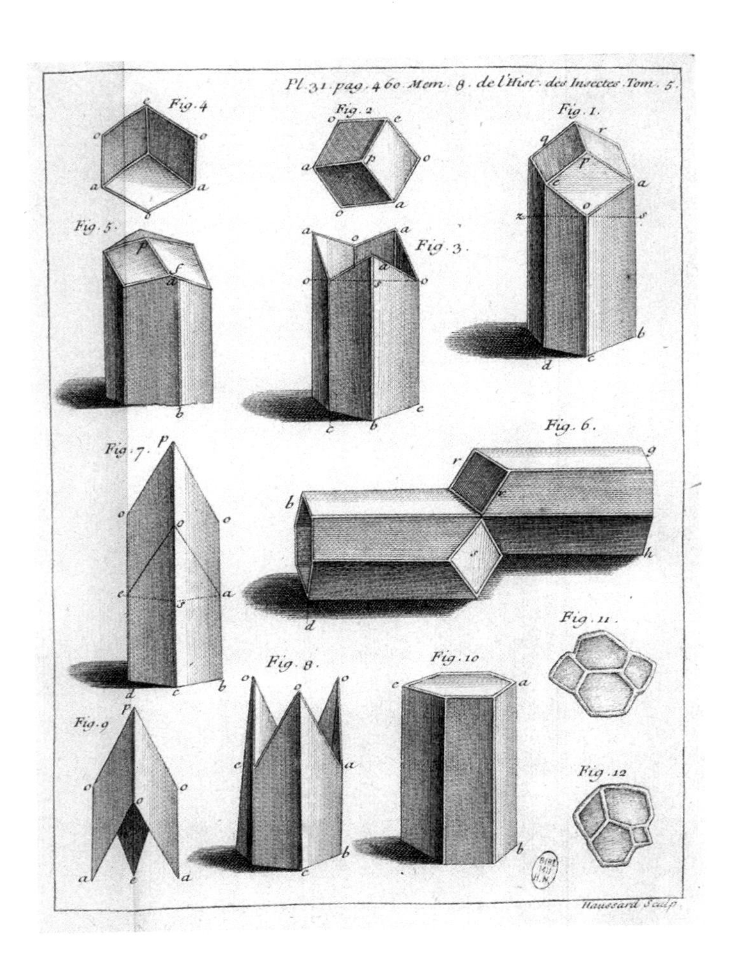

Though beekeeping and beekeepers are constantly changing, honeybees are still doing what they have always done, to the extent they are allowed to, that is. Their hexagonal wax cells cannot be improved upon. The greatest possible volume with the least possible consumption of material. Image from the fifth volume of Réaumur's Mémoires pour servir à l'histoire des insectes *of 1750.*

Apart from various digressions, the first part of this book deals with the history of bees and beekeeping, both in the recent and the more distant – occasionally very distant – past. Part Two takes place in the present, which is always a great deal harder to comprehend. The bigger picture can be made out more or less: global financial interests – the chemical industry and agribusiness – on the one hand; on the other, the bees and other pollinators whose survival they threaten. Who is winning? Robot bees that can survive anything are already in production.

Down at hive level the situation appears to be too diverse for an overall view. There are professional, amateur, traditional and alternative beekeepers, queen bee farmers, bee breeders, tree beekeepers, activist beekeepers, migrant beekeepers and urban beekeepers. Some swear by the Buckfast bee, others by the Carnica bee or the dark or yellow bee. Some advocate one model of hive, others another, or a fifth model or a tenth. And even though *Apis mellifera* is the animal that most scientific research has been devoted to, science is constantly making new discoveries. Striking similarities have been found between the way a human brain and a bee swarm make decisions.

I hope that these reports from a few of my excursions into the beekeeping world in recent years will provide an impression at least of what a varied place it is.

A beekeeper association meeting at Ultuna in August 1948. How different these authoritative gentlemen appear to the casually dressed audience at BEECOME 2016, which at least as many women as men attended. In the past the wives did a lot of the bee-related tasks at home, but when it came to the affairs of the association, either they did not attend or they just made the coffee. It was not until the end of the twentieth century that "beekeeper" ceased to be an exclusively male title. The hives have changed as well. Wooden tray hives are a model seen only rarely today. Now it is stackable hives made of expanded polystyrene that are in vogue.

BEE WELCOME, OR NOT

BEECOME, THE ANNUAL EUROPEAN congress of beekeepers, is taking place in Sweden for the first time. The theme is "Beekeeping for a Sustainable Future", the place is the conference centre Malmö Live, the year 2016. The organisers are the Swedish Professional Beekeepers, the Swedish Beekeepers Association, the European Professional Beekeepers Association and the Swedish University of Agricultural Sciences. Scientists, celebrity chefs, representatives of Swedish and international beekeeping organisations, and of public authorities and the E.U., plus busloads of professional and amateur beekeepers are taking part – actively or passively or both – for the three days of the programme, which is loaded with debates, talks and workshops on everything from combating the deadly Varroa mite and the effects of climate change on beekeeping to mead production and hive construction.

Like more or less everything else, the beekeeping world has changed radically since the time I kept bees of my own. Congresses with thousands of visitors and international guests would have been as inconceivable then as swimming competitions on the moon. Beekeeping was a male-dominated and fairly provincial affair in those days, in organisational terms at least. The Swedish Beekeepers Association has now elected its first female president and 40 per cent of new beekeepers are younger women. I can't help thinking about my former acquaintances at S.S.B.F., the Beekeepers Association of Southern Sweden, and wondering what they would have thought of this event. Presumably that the admission charge was too high for them to even bother registering, and in any case they already knew everything a beekeeper needs to know.

There are exhibitors here from near and far. T.A.B., the Turkish Beekeepers' Association, is advertising an autumn seminar on pine honey on the beautiful Turquoise coast. There are tea services decorated with bees for sale from England, chemical preparations from Italy against bee diseases and parasites, and beauty products made of *gelée royale* from France. You can order wooden hives from Estonia, expanded polystyrene ones from Britain and queen bees from Malta.

A Varroa mite hugely magnified. Wow! You can order counteragents here such as oxalic acid, lactic acid, formic acid and thymol. "When a bee colony has been attacked by Varroa mites, the infection must be combatted or else the colony will die," according to the Swedish Ministry of Agriculture. Other people, however, think that if you constantly intervene to combat the Varroa mite, bees will never develop a natural resistance, and untreated bees have in fact become resistant on occasion.

The pastel shades of the protective suits from England seem to be in great demand, while many people are drawn to the mead samples. It couldn't be more apparent that beekeeping creates business opportunities galore.

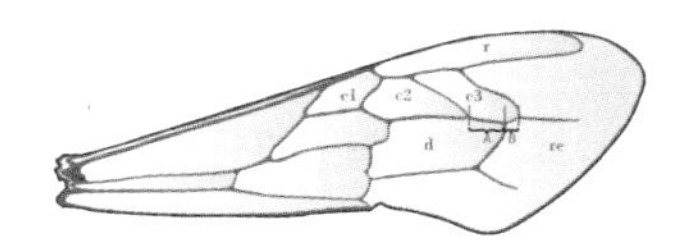

Wing morphometry is considered a good way of determining what race a bee belongs to, and how pure-bred it is. This is a wing from a European Dark Bee.

But there are idealists here as well, who are not attending the congress to make money. I get caught up at a stand providing information about *Apis mellifera mellifera*, the original European bee, known here in Sweden as the black Nordic bee. It is at this stand that Ingvar Arvidsson tells me that until the end of the nineteenth century this was the only bee kept in northern and Central Europe, from Ireland to the Urals and from the Alps to central Scandinavia. But it has been on the retreat ever since the yellow Italian bee, *Apis mellifera ligustica,* and the Carnica bee, *Apis mellifera carnica*, were introduced to the more northern countries during the nineteenth century. By the time Brother Adam's Buckfast bee began its triumphal progress in the 1970s, *Apis mellifera mellifera* was considered more or less extinct.

That would only change at the last possible moment. In Sweden a few enthusiasts including Ingvar Arvidsson launched Projekt Nordbi (Project Nordic Bee) in 1990 and then Nordbiföreningen (The Nordic Bee Association) in 1997. The first priority was to find out how many colonies of dark bees were left in Sweden. Appeals for information were sent out and happily it transpired that were still five hundred more or less pure-bred colonies in rural provinces such as Jämtland, Västerbotten and Dalsland. The queens selected from the best of these were used as breeding stock.

There was increasing interest in the original bee in other countries as well: survivors were eventually found and efforts were made to propagate the strain. In 1995, an international cooperative organisation (S.I.C.A.M.M.) was formed that encompassed members from sixteen European countries, from Ireland in the west to Russia in the east.

What are the characteristics of the Nordic variety of *Apis mellifera mellifera*?

"It is a robust bee that has adapted to our climate and our flora and is long-lived," Ingvar Andersson responds enthusiastically. "It is better able to cope with the winters than other bees, it doesn't require a great deal of winter feed and will gather nectar even when it is cold and windy."

A passer-by stops to listen.

"The dark bee isn't worth keeping!" the woman says angrily. "It's so bloody irritable!"

He has heard this kind of thing before.

"That's what a lot of people think even though it's a very placid bee, but it acquired a bad reputation when they started importing other bee species. When you cross the dark bee with other varieties, the result can be extremely aggressive hybrid colonies."

Even though it feels rather mean, I ask whether people confused the campaign for the Nordic bee with far-right extremist organisations such as Nordfront and the Nordic Resistance Movement.

Ingvar Arvidsson sighs.

"Yes they do, but that's a complete misunderstanding. Projekt Nordbi and Nordbiföreningen are in no way nationalistic organisations; on the contrary, we are in favour of diversity and that is precisely why we want the Nordic bee to survive."

I wish him luck and continue my tour. There is more to discover. Like the fact that in the southern Swedish town of Höör they run a two-year correspondence course for professional beekeepers; that the Svenska Bin (Swedish

Bees) association administers a project called Bee Welcome that encourages beekeepers all over the country to provide newly arrived refugees with opportunities to work and train, and that the E.U. has a reference laboratory for bee health in Antibes in France.

While there are no actual bees here, there is a vast array of bee products, and not just honey. Propolis is the fragrant putty bees produce from sap, resin, saliva and wax to seal cracks and holes in the hive, and which keeps bacteria and other unwelcome guests away. John Larsson used to scrape off a bit to chew on. He said it was to prevent him getting colds.

But what is that?

"Please, have a taste," the saleswoman says, holding out a dish of tiny brown-striped nuggets.

I hesitate. Only they are not what they look like, but a fermented blend of pollen and nectar called "bee bread", which is supposed to contain large amounts of amino acids, vitamins and other healthy substances. Bees use it to feed their worker and drone larvae. At the beginning of the twenty-first century a method was invented in Latvia to harvest it in large quantities, and it is now being marketed as a health food for humans as well.

Bees nourish their larvae with "bee bread", tightly packed pollen mixed with nectar and digestive fluids, and sealed with a drop of honey. It is also consumed by people as a health food.

"Does it make you as busy as a bee?" a facetious gentleman asks. "Or as lazy as a drone?"

I am nibbling at a bit of bee bread that tastes slightly acidic. Why weren't the bees allowed to keep it? Their brood need it much more than we do.

Among the subjects addressed in the concluding panel debate are how farmers and beekeepers

In England, Germany and France, beekeepers have staged major demonstrations against the use of agricultural poisons that kill bees and other pollinators, and against the companies that produce them. In 2018 the E.U. banned the three most dangerous neonicotinoids, but there are still several products on the market that are toxic to bees.

can work together, and what political decisions need to be taken. The issue of the toxicity of neonicotinoids obviously comes up. In 2013 the E.U. introduced a temporary ban on their use in rape farming. Jytte Guteland, a member of the European parliament's environment, public health and food safety committee, says that a further tightening of the rules is essential to protect bees and the ecosystem.

She is opposed by Julian Little, spokesman for the chemical giant and neonicotinoid producer Bayer's own "Bee Care" programme, whose researchers have discovered that it is only in a laboratory environment that bees are harmed by neonicotinoids. Scientists who do not have links to the chemical industry have arrived at very different conclusions.

Geoff Williams from Canada, for example, reported at this very congress that they not only harm worker bees, but also lead to the queens being less able to lay eggs and living shorter lives. But Dr Little insists that we have to deal with the real world:

"What are the consequences of banning one product one moment and another the next? We could of course tell farmers they should stop growing crops or just grow them on a limited acreage. The problem is that in that case Europe would become entirely dependent on imports from other countries to feed our populations, and that is unacceptable."

Marie-Pierre Chauzat from the European Union reference laboratory for bee health demands the right to reply.

"Farmers managed without toxins in the past, and they could do so now too. We have to work with nature and not against it."

Applause from the audience. Then Lasse Hellander, committee member of the Swedish Beekeepers Association and an ecological farmer and beekeeper, is given the floor:

"This is not just about the chemical industry and farmers," he says, "but about consumers as well and how much they are prepared to pay for non-toxic products."

Money, money, money. A little, a lot, vast amounts of money even. An event of this kind can leave you with the impression that all beekeeping is about is money.

But it isn't.

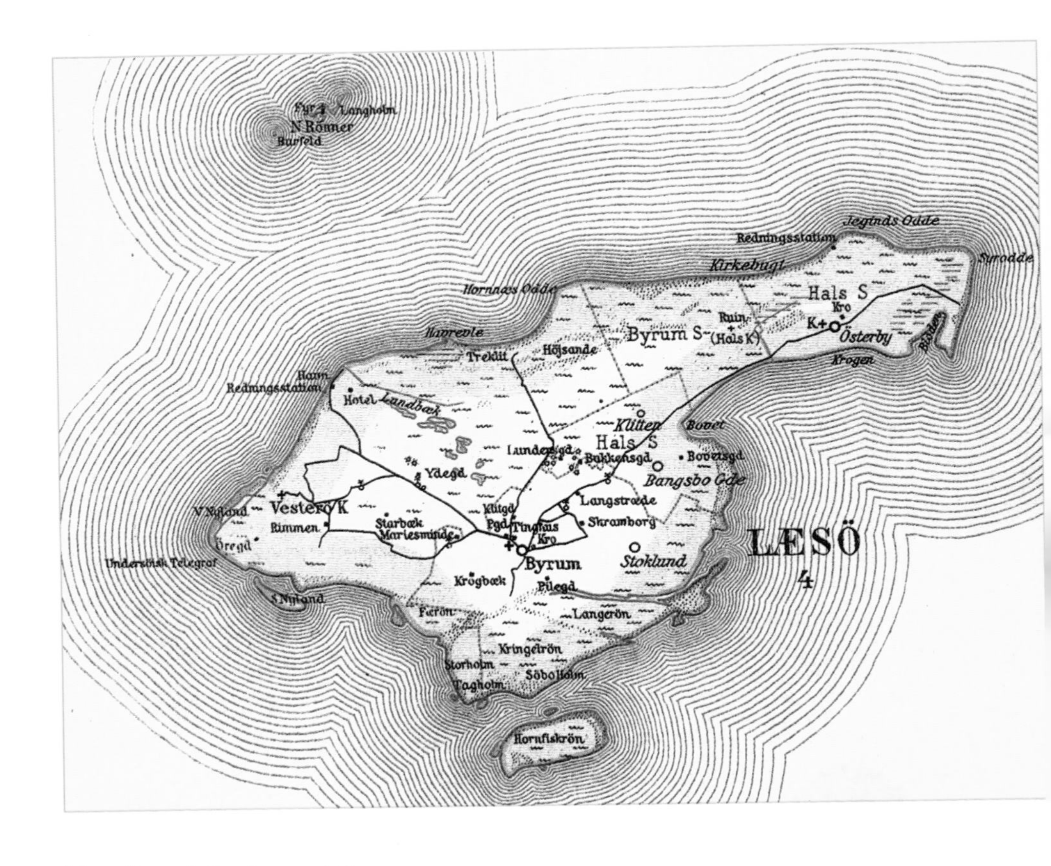

Map of Læsø dating from c. 1900

THE BEE WAR ON LÆSØ

LÆSØ IS A DANISH ISLAND in the Kattegatt, halfway between Frederikshavn and Göteborg; much the same size as Jersey, it has just under two thousand permanent residents. The island is known for its scampi, its salt works, the traditional eelgrass thatching on some of its houses (not many of these are left), its sandy beaches and because the artists Per Kirkeby and Asger Jorn lived and worked here. Læsø is also known for its colonies of European brown or dark bees, referred to as black bees in some countries.

The first time I heard about them was when Ingvar Arvidsson, whom I met at BEECOME, told me that a particularly pure race of *Apis mellifera mellifera*, the original European bee, had been found on Læsø and that the island is the only place in Denmark where this bee has survived. When a law was passed that banned other races than the dark bee from the island, the beekeepers there who kept yellow Italian bees were furious. A veritable civil war broke out and it lasted for many years.

Ingvar Arvidsson did not seem like someone given to exaggeration, but could that really be true? When I got home

I Googled "the bee war on Læsø" and discovered a wealth of articles and investigations, a Wikipedia page, legislative proposals, police records, parliamentary debates and court case verdicts. Taken together they built up a picture of a chain of events that warrants careful consideration by anyone who believes beekeepers are peaceful and well-intentioned people. Ingvar Andersson had not been exaggerating, quite the contrary.

Not that I like war in any form, but I had finally found a reason to travel to Læsø, a place I had long dreamed of visiting. First, though, I had to read up on the subject. Here is a summary of what I found:

In 1983 local headmaster Alfred Petersen applied to the government's Bee Disease Committee to declare Læsø a breeding sanctuary for the European dark bee, the only bee that had existed on the island until the 1970s, when several beekeepers introduced the yellow Italian bee, *Apis mellifera ligustica*. The committee approved Petersen's application, but the decision was completely ignored by the owners of the yellow bees. No-one was going to tell them what kind of bees they could have.

Now, some of you may be wondering what could be so troubling about yellow and brown bees being allowed to cohabit? There were, after all, no negative feelings between the two races. And that was the problem: brown queens would happily allow yellow drones to mate with them. Were cross-fertilisation allowed to continue, in the end there would be only hybrid bees on Læsø, and a valuable source of genetic material would be gone for ever. That was the view taken by the keepers on the brown side and the scientists who specialised in bees.

On the yellow side it was honey production and profitability that counted. Because the yellow bees produced more honey than the browns, they were going to go on keeping them and they had nothing against their queens coupling with brown drones. Hybridised colonies appeared to be extra hard-working whereas, according to its opponents, the dark bee was sickly and produced less. As time passed, the distrust between the brown and yellow beekeepers grew. They stopped talking to each other, which is hardly something that goes unnoticed on a small island. Quite a few of them were related to the other side and things could get pretty awkward at gatherings such as birthday parties, confirmations, weddings and funerals. The yellows formed the Free Beekeepers Association, while the brown side was organised under the name of the Læsø's Beekeepers Association, which was a subdivision of the national Danish Beekeepers' Association. The Establishment, that is. While scientists were tracking the genetic changes among the dark bees, it is surely a matter of considerable regret that no anthropologist or social psychologist studied the changing attitudes among the beekeepers of the two factions and what factors helped make the conflict worse. Why did it become "us against them" and what determined which side you ended up on? Questions that are endlessly repeated throughout the history of the human race, in matters both large and small.

Things continued in this way until the state intervened in 1993 as an interested party. The Minister of Agriculture and Fisheries in the then Social Democratic government issued a draft bill that was in line more or less with Alfred Petersen's application of ten years previously. In accordance with the goals set out in the Rio Convention to assure the diversity of flora and fauna, the only bee that would be allowed on Læsø was *Apis mellifera mellifera*. Those with yellow bees would either have to move them off the island, exchange their queens for brown ones or destroy their colonies, in which case they would be fully compensated for any financial loss.

The Progress Party, Mogens Glistrup's tiny anarcho-populist party and the forerunner of the Danish People's Party, protested vehemently against the decision. The Progress Party had made a name for itself by rejecting taxes in general and the then generous national policy on immigration. It demanded a total ban on migrants from Muslim countries and that those who had already arrived should be sent home. But when it came to the bees on Læsø the Progress Party raised no objections to immigration and the mixing of the races. Quite the contrary. If the original brown bees were to become the sole species on the island once more, that was bound to lead to inbreeding. The bee population was too small and introducing genes from outside could only have a positive effect.

The law was passed by a clear majority in the Danish parliament, however, and the war, which had been merely smouldering until then, burst into flame. Even before the legislation came into force a number of yellow bee colonies were vandalised. Some of the yellow beekeepers who refused to comply with the ban had their breeding stock confiscated. Ditlev Bluhme, one of the yellow generals in the bee war, was prosecuted at the criminal court in Frederikshavn for breaking the law by continuing to keep his bees on the island. His legal team maintained that the ban against yellow bees was in conflict with the provisions of the E.U. treaty on the free movement of goods and on the conditions for trade in thoroughbred animals. He was convicted but appealed.

After many twists and turns the case finally arrived at the E.U. Court in Luxembourg. The verdict was that while the Danish law constituted a restraint on trade, this was outweighed by the threat to a species of bees. 1–0 to the browns.

The yellow side came up with a different interpretation: "Our argument that the Læsø Law was in conflict with the Treaty of Rome which forbids restraints on trade was borne out. The judgment was a verdict critical of the way the matter

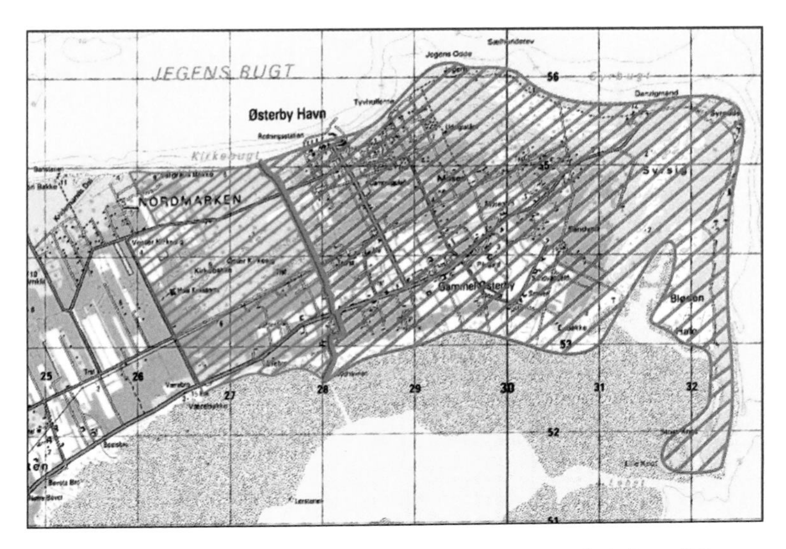

Læsø's north-eastern cape. Within the red cross-checked area only Apis mellifera mellifera, *the European dark or brown bee, may be found during the mating season. To the west is a buffer zone whose purpose is to stop brown queens from breeding with the drones of other subspecies.*

has been dealt with by the authorities that have fought for many years to stop us bringing it before the E.U. Court," declared the lawyer for Ditlev Bluhme who demanded that the E.U. Court should reconsider the case on the basis, this time, that the law should have been put to the E.U. before it was voted on by the Danish parliament.

Things went on in this vein for another couple of years while the risk grew ever greater that the brown bees would be so hybridised with the yellows there would be nothing left to preserve. In 2005 a study showed that only 25 per cent of Læsø's bees were still predominantly brown. Not thoroughbred, that is, but with mostly brown genes.

But at the very last minute the then Conservative government's food ministry signed an agreement with the beekeeper associations on Læsø which ensured

Christian Juul and a hive containing a swarm he has just captured. The bees were a hybrid of the yellow and the brown, but that does not worry him. "They're usually very productive," he says.

that the eastern cape of the island was divided into zones that – after a few adjustments – are still in force today. A Cypriot solution. Only brown bees are allowed in the easternmost zone during the mating season. To the west of this brown zone is a buffer zone of six kilometers, while yellow bees and hybrids are free to roam over the rest of the island. According to the follow-up surveys that have been carried out, the system seems to work. As long as there are sufficient suitors in the brown neighbourhood, no virgin queen is going to make the long trek west in order to be fertilised.

Twelve years have passed since the island was divided, and I am here to try and find out how things have developed between the brown and yellow beekeepers in the interim. I have been informed by Carl-Johan Junge, the chairman of the browns, that he holds a lecture every Sunday in July at

Skovhytten, a forest lodge in the Klitplantage nature reserve. All I have to do is turn up. As for the yellows, bee researcher Per Kryger, who has been monitoring developments on Læsø, recommended I get in touch with Christian Juul. I only managed to get hold of him once I was on the island, but that turns out not to be a problem.

"I've been a bit poorly," he says over the telephone, "but I'm back on my feet again now, so just come round this afternoon." He lives on a farm with the lovely address Vestre Himmerigsvej (Western Heaven Road). The very first thing he does is apologise for not being fully fit, but that is because of the medicine he has to take.

"I'm beginning to get old," he says. Though he does not look particularly old and it turns out we were born in the same year. Somehow that gives us a sense of shared identity, illusory, of course, but it serves as a kind of bond. I had been imagining the yellow beekeepers as a gang of confrontational and unpleasant characters, but Christian Juul is not like that at all. He is a quietly spoken and kindly man and when he talks about his life I think I can understand why he is on the yellow side. He is a Læsø-ite of the old sort.

This has always been a poor island, and the islanders have needed many different sources of income. Wrecking was the best way of earning money in the past; now it is tourism, a more sophisticated form of plunder. Christian Juul has done more or less everything. He has collected eelgrass for thatching, fished for shrimp, scampi and eels, been a gravedigger, kept his farm going, raised pigs and run a farm shop together with his sons. He and his wife Amy have also kept bees on a large scale.

"Though we can't do as much now as we used to," he says.

The farm shop has been sold and the pigs butchered. The barn is full of beekeeping equipment although he only has a few colonies left. He opens a hive that contains a swarm a neighbour asked him to capture. The bees inside are definitely not that yellow. "That's how it is now," he says, "most of the colonies are mixed [. . .] not that it matters. The hybrids are good at producing honey."

We are sitting in the sunshine by a pond that Christian Juul dug out in order to farm fish, yet another means of making money. When we eventually get onto the subject of brown bees he tells me that Amy needs to be part of the conversation. She finds sitting outside difficult owing to a skin problem, so we go inside to the kitchen where she is doing the crossword and drinking coffee.

"It keeps my brain ticking over," she says.

Christian brings us beers, as the Danes do when people visit, and Amy tells me that she comes from the capital Copenhagen and therefore tends to look at things slightly differently from other people on the island, and that applies to the owners of the brown bees as well. Christian clears his throat and says that although they're decent people on the whole, he doesn't like their bees. They're not as hygienic as the yellows: they don't remove their dead comrades from the hive, and they're not particularly productive. You can't make any money keeping brown bees.

Amy interrupts. "You're being too nice!" It's not just the bees there's something wrong with, it's the people who keep them as well. Not to mention the fact that they get five hundred Danish crowns from the government for every overwintered colony if the queen has mated within the reserved zone. That's not right, she says. No-one checks that they really have got as many colonies as they say, and she knows that there's been cheating, indeed she does.

"And then they try and get people to believe that keeping brown bees is somehow better, and why should that be? Someone coming along telling me what to think makes me suspicious. This business with climate change, for example,

Carl-Johan Junge, chairman of the association of brown beekeepers on Læsø, giving a lecture on bees in general and Læsø's bees in particular.

and the idea that it's the fault of the human race, I don't believe that for a moment. Even though the earth's climate has been constantly changing, we're supposed to feel responsible for the way things are now. It's the same with the bees. We're supposed to feel ashamed we don't want to have brown bees, but we're not!"

"Their methods haven't always been fair," Christian concedes. "For a while they tried to bribe me over to their side. But they have been good at the P.R. side of things, at making a name for their cause and for their honey, even though it's exactly the same honey as you get from the yellow bees. After all, it's where the bees get the nectar from that determines what the honey is like, not the species of bee."

Then we go into the living room to look at the pictures Christian has painted. Sky, sea, clouds, sunlight, moonlight.

The next day is Thursday and I take the bus – which is

free here on Læsø – to Byrum in order to walk from there to Skovhytten, where Carl-Johan Junge is giving his talk. Unfortunately I didn't take a careful enough look at the map. The walk takes almost two hours along deserted roads; even so, I manage to arrive just in time for the lecture. The audience is made up of summer visitors, and it is obvious that Junge is a seasoned lecturer. After an account of the way a bee colony is structured, he picks up a drone and squeezes its rear body so its tiny and transparent sexual organ protrudes. Then he walks among the audience showing the bee's parts to his listeners, even though they are almost impossible to see. Interesting, but slightly odd.

After the lecture I ask him if we could have a talk, but it turns out there are some matters the committee of the beekeepers' association need to discuss before he can get back to me.

The audience has left, and Skovhytten soon empties apart from the committee members who are sitting around a table beside the parking lot. It is humid and there is a rustling in the trees, maybe a thunderstorm is on its way? It is a little disappointing to have to wait, but what is worse is that I am so dreadfully thirsty. I didn't load up with water before my long walk here, and now I am hunting around for a tap. There does not appear to be any such facility in the forest lodge. On the other hand there is an exhibition about bees, which I study closely to make the time pass. In the end I feel bold enough to go up to the gentlemen on the committee, who are drinking beer.

"We've almost finished," Carl-Johan Junge says, "you can have a seat if you're feeling particularly brave."

Then he asks if I'd like a beer. I would. Very much!

"Oh, sorry, there aren't any left," he says.

I ask – with considerable restraint – if I could possibly get some water and, would you believe it, the committee secretary goes and fetches a bottle from his car.

Carl-Johan Junge asks me what it was I wanted to talk to him about. So I ask if there will ever be peace between the yellow and the brown beekeepers?

"No," he says, very deliberately. "The yellows have been a thorn in our side for decades. Besides the law cases, they have kept mucking us around in all sorts of ways, like printing labels that are deliberately meant to be confused with the ones the browns use. Though they'll all be dead soon, which is something I'm looking forward to."

"Even if we're starting to get old ourselves," says the man who got me the water.

The others nod. That's true, all they can do is hope that the few new members they have will grow in number and can take over the work of conservation. "There is only one sensible solution," Carl-Johan Junge says, "and that is for the whole island to be reserved for the brown bees. But that would require all the beekeepers to agree, and that's not going to happen while we're still alive. Or the political will to get it done, but that doesn't seem to exist either."

Why do I never learn, I ask myself when I get back to my hotel in Vesterhavn, driven there this time by the nice secretary, that everything is so much more complicated than I imagine?

In 2016 a nationwide Forening for Brune Bier (Association for Brown Bees) was set up in Denmark which means that if *Apis mellifera mellifera* is to have a future in the country it is no longer solely dependent on the beekeepers of Læsø. Another island, Endelave, has been entirely reserved for the brown bee and, unlike Læsø, doesn't have a faction among its population that stubbornly insist on the right to decide for themselves what kind of bees they are going to keep.

This painting in the cave system at Cuevas de la Araña near Valencia is some 8,000 years old. A man or a woman has climbed a rope ladder or a tree to gather honey from wild bees.

NATURAL

OR JUST NATURAL-SEEMING BEEKEEPING?

Owing to their free and natural lives, forest bees are less troubled by ailments and discomforts than domesticated bees.

Carl Hårleman, 1749

EVEN THOUGH MUCH IS MADE nowadays of natural beekeeping, is there really any such thing? The whole of agriculture is a departure from nature, after all. Only wild bees live natural lives. If you are going to be picky there is no such thing as natural beekeeping, just varying degrees of unnaturalness. Or of seeming natural.

Why pay attention to words, though? Natural beekeeping is an umbrella concept that covers various methods of keeping bees that all have in common the beekeeper putting the health and welfare of the bees before profitability. And if a bit of honey gets produced along the way, that is all to the good, of course.

But it is not just an umbrella concept; it's also a wasps' nest. The beekeeping establishment and its various associations start pointing their stings whenever this kind of "mumbo jumbo" is mentioned, and I am writing this chapter with some

Tree beekeeping was common in Sweden well into the eighteenth century, and until even later in some countries; in Bashkiria, for example, the tradition is still alive. Hollow spaces were carved into the tree trunk some six meters above the ground; lids with holes were then fitted over them so the bees could fly in and out. You either waited until a swarm found their way in or you moved a swarm into the hollow. Is tree beekeeping more natural than using hives? In some ways, it is. When bees are allowed to choose for themselves where to settle it is always high above the ground, and the round walls suit their construction methods better than the rectangles of the modern hive. This picture is of a Polish tree beekeeper – a bartnik. *Today there are* bartniki *once more who have learned their skills in Bashkiria and are handing on the knowledge to students in other countries such as Germany and England.*

trepidation. I risk becoming a target, particularly as I find the methods in question interesting and worth taking seriously even though I have no experience of my own when it comes to natural beekeeping. But if I'm going to write about the contemporary beekeeping world I can hardly pretend the phenomenon does not exist. Still, nothing daunted, nothing gained!

In the bee world of yesterday, in which I played a minimal part, though a part nonetheless, there was only one way to keep bees. Even if my guru John Larsson was part of the opposition to the National Association of Swedish Beekeepers, this did not apply to the basic principles of beekeeping. So he taught me to treat my bees like everyone else did, in accordance, that is, with the principles of so-called modern beekeeping. These had been in force since the nineteenth century when a couple of revolutionary inventions changed the lives not only of beekeepers, but even more radically of the bees themselves.

The American priest Lorenzo Langstroth's revolutionary hive with moveable rectangular frames. All very neat and tidy, though not for the bees' sake but for that of the beekeeper who could now work more efficiently and on a larger scale than could have been conceived of before.

The most important of these were the new hives with moveable rectangular frames fitted with prefabricated wax foundations. These prevented the bees from constructing their combs with cells in various sizes for different purposes. The foundations meant that all the cells were the same size. All very neat and tidy. Among the benefits of the new moveable frames was the fact that they could be accessed and adjusted time and again, unlike the old straw hives and the even more ancient log hives in which the bees were left to fend for themselves until it was time to harvest the honey.

Another innovation was the extractor, which made it possible to remove the maximum amount of honey from the frames. To that should be added the falling price of sugar, which made it profitable to take all the honey the bees had produced and replace it with a sugar solution the bees could feed on during the winter. The breeding of queen bees and the trade in queens would also change the nature of beekeeping. A colony could now be given a new queen of the best pedigree that had been fertilised, if so desired, with the appropriate genes before the rate of the old queen's egg-laying began to diminish.

Taken together this meant that beekeeping could be run on a much larger scale, and honey became a far more profitable commodity. Out went straw and round woven hives; in came wooden (and subsequently expanded polystyrene) hives, consisting of rectangular boxes with frames that could be stacked one on top of the other when new space was required, or which were capable of being adapted in other ways. Although the equipment has been modernised and new gadgets invented, there have been no major changes in the way bees are kept since that time, apart from in one area. There are now numerous toxic substances available to treat diseases and parasites, which occur more frequently than they used to.

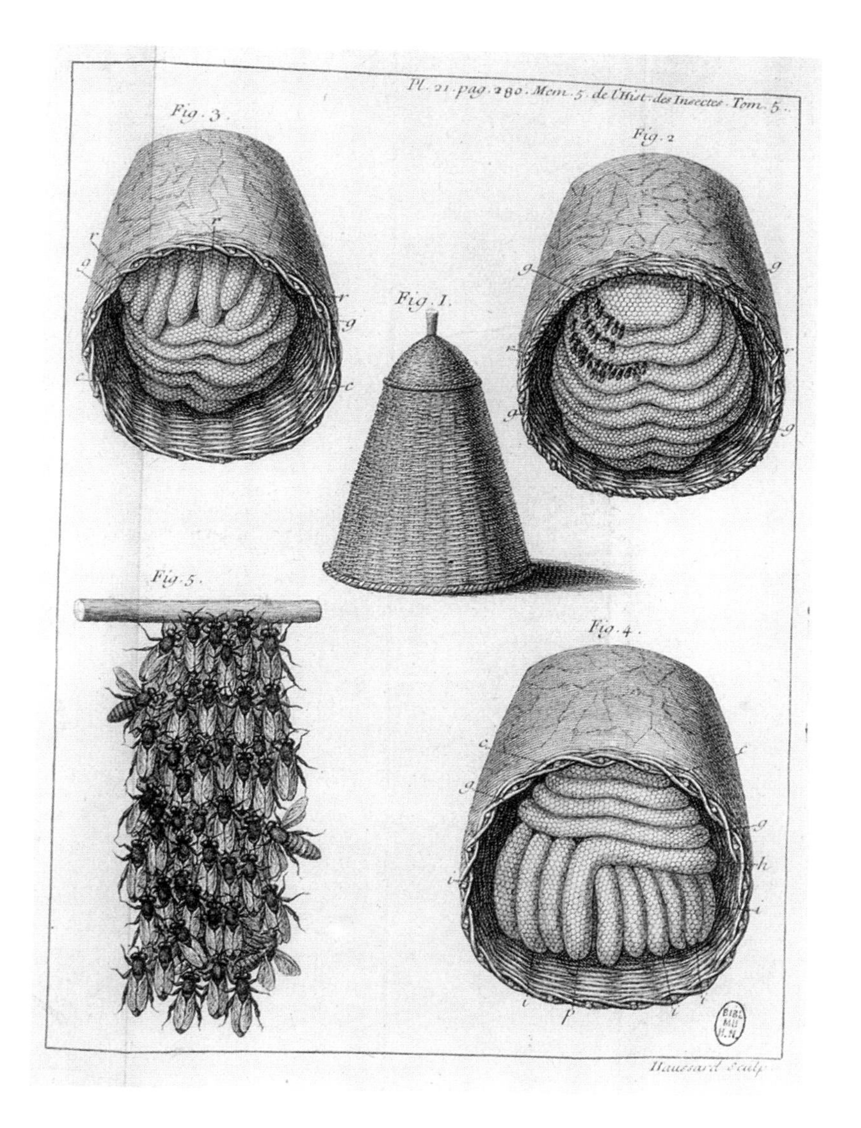

It is natural for bees to build their combs in a space that is round like the inside of a hollow tree. They could do this in the old woven hives. Although no one construction was identical to another, none of them were rectangular. The bees have just begun building a comb (bottom left) that is fixed to a stick or bar at the top of the hive. From Réaumur's Mémoires pour servir à l'histoire des insectes.

A beekeepers' meeting in Hallsberg, Sweden, in 1909. Perhaps these solemn gentlemen were discussing the advantages and disadvantages of the modern frame hive as compared to the old straw hive.

This photograph was taken in Bohuslän, Sweden, around 1930. It did not take long, however, before all beekeepers were convinced of the excellence of the frame hive.

Not everyone was on board with the new methods, however. Cautionary voices were raised that they failed to pay enough attention to the needs and instincts of the bees themselves. Lars Clementz Svensson in Vallkärra championed what he called rational natural beekeeping: "This senseless dividing of the bee colony and fiddling with it will not go unpunished. The bee follows and must follow the laws that nature and age-long experience have etched into its awareness."

Such objections were rare, however, and would soon disappear from the debate entirely. For most of the twentieth century, beekeeping in Sweden was dominated by the national association – with modernity and rational beekeeping in the driving seat. You had to go abroad to find examples of a different way of thinking.

In 1946 the Austrian priest Johann Thür described how the particular heat and marvellous scent of the air in the natural hive was crucial to the bees' health and well-being, all the more so as they discouraged harmful bacteria and other undesirables. When you open the hive and lift out the frames for inspection, this *Nestduftwärme* (the hive's scent and heat) disappears and the bees have to spend a lot of their energy on recreating it.

The Frenchman Émile Warré (1867–1951) – yet another priest – tested more than three hundred different models of hives before he constructed his own: *la ruche populaire* (the people's hive). He, too, thought that the hive's scent and heat were crucial and his hive was therefore intended to be opened as infrequently as possible. He also felt it was vital that a beekeeper should not have to invest in expensive equipment. Everyone should be able to afford to keep bees. Like countless priests before him he pointed out the beneficial moral effect of bees. They demonstrated the importance of work, order and devotion to the common good, while also

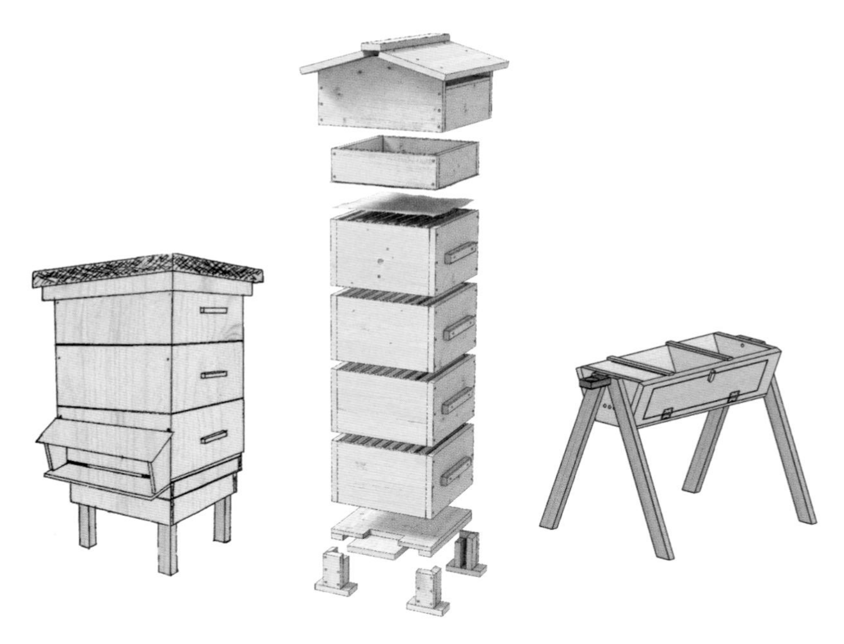

From left: a stackable hive (a common model nowadays), a Warré hive and a top-bar hive. New boxes are added to the stackable hive from the top, and to the Warré hive from below.

keeping the beekeeper away from the pub and other sources of corruption, the promiscuity of modern sport in particular. In Sweden, however, until the end of the twentieth century it was only in anthroposophical circles and under the influence of Rudolf Steiner's ideas that people understood the methods that were good for maximum honey production were not necessarily good for the health of the bees.

The boom in alternative approaches would arrive only as a result of the increasingly alarming reports on the threat to the survival of bees as a species. Could the decline of bees be the result not only of toxins, the agricultural emphasis on monocultures, and the Varroa mite, but of modern beekeeping as well?

A whole range of new kinds of bee dwellings quickly appeared, though several of them were in fact ancient. A daring young Pole, Piotr Piłasiewicz, set off for Bashkiria in Russia, where bees were still kept in hollow or hollowed-out tree trunks, and then reintroduced tree beekeeping to his homeland. From there it has spread to other countries, and there are now courses on how to hollow out a log for the purpose. There are barely any hollow trees to be found in the wild now. In the Cévennes in southern France, the tradition of using hollow logs from the wood of the sweet chestnut was revived, as was that of the characteristic straw hives of the Lüneburg heath in Germany.

Warré's "people's hive" was introduced to the English-speaking world in the book *Natural Beekeeping with the Warré Hive* by David Heaf, which led to it becoming a kind of cult hive outside the French-speaking world as well. The horizontal top-bar hive, a model the Greeks were already using in antiquity, has become even more popular. It was reinvented in African countries where hives were needed that were cheap to build and easy to use. The top-bar hive has only one layer with bars on which the bees build their combs as and when they see fit.

The top-bar hive as featured in Philip Chandler's *The Barefoot Beekeeper* has also become a favourite with British beekeepers of the new school who want a couple of easily maintained hives in the garden for the sake of pollination, to produce a bit of honey for private consumption, and because it is fun. When the celebrated gardening writer Alys Fowler acquired a top-bar hive and wrote a book about it, its success was guaranteed, particularly as it was also demonstrated on television.

A rather peculiar hive is the invention of the German anthroposophist sculptor Günther Mancke: the *Weißenseifener Hängekorb,* better known as the Sun Hive. Mancke got the idea

An entirely new variant: the elliptical Sun Hive, inspired by the combs that wild bees construct. It can be covered in cow dung to provide insulation.

for it while studying the combs of honeybees that had gone wild. The hive is egg-shaped and made up of two straw hives with a wooden board between them and is caked with cow dung (organic, of course) to insulate the hive, keeping it cool in summer and warm in winter. It is meant to be hung at least two meters up in the air because the places bees themselves choose to settle are never close to the ground.

But alternative beekeeping is not just about hive models. You can keep bees the natural way in an ordinary stackable hive and the conventional way in the top-bar version. Of far greater importance is the way the bees are treated. The main and most controversial difference between the two camps is winter feeding. Alternative beekeepers allow their bees to overwinter on their own honey, preferably on its own but not infrequently mixed with sugar. Because bees have been living off their own honey since the age of the dinosaurs and have managed perfectly well, they should be able to do so now too, particularly as honey contains substances that form a natural protection from undesirable microorganisms and diseases.

Statements like these may be dismissed with outrage and indignation by the beekeeping establishment. No scientific foundation, just opinions, inaccurate data! Research and tried and tested experience also prove that ordinary sugar is excellent winter fodder, whereas honey can even be harmful.

Why all the commotion? The answer may be found as far back as 1907, in a letter sent in to *Bitidningen* (The Bee Magazine) by J. Enelund in Hjälstaby.

> “Experience has shown that sugar is an excellent winter food for bees. If it is provided in the appropriate way and at the right time, it remains fluid in the combs and palatable to the bees, which is not always the case with honey [. . .] Because sugar is much cheaper than honey there is an immediate financial gain to be made from exchanging honey for sugar.”

The matter would not be described so crassly nowadays, but it goes without saying that not all beekeepers are prepared to abstain voluntarily from the majority of the twenty kilos of honey per colony their bees produce to live off over the winter.

Other common points of dissension are ready-made foundation sheets of wax or plastic, the importation of bees, and queen breeding. There is, however, one area in which natural beekeepers do not differ from the others: they form factions. There is always something to disagree about. The best-organised part of the movement is made up of the anthroposophically inspired. Even so, the outspoken champion of the top-bar hive, Philip Chandler, has no patience for Steiner's notions about bees.

> “What is perhaps more surprising – and infinitely less helpful – is when people concoct mystical ‘explanations’ derived entirely from their imaginations and pass them on as if they had some scientific validity. Possibly the worst offender in this respect in modern times is Rudolf Steiner, whose flights of fancy concerning bees bear not the slightest resemblance to observable reality, yet are

VIVECA, A WARRÉ HIVE BEEKEEPER
In order to stress the bees as little as possible, Viveca only opens the top of her hives to harvest any left-over honey in the spring. So no summer harvest at all. New boxes with bars to which the bees can attach their combs are added from below. This allows that vital hive scent and heat to be retained, and the bees are able to build from the top down, which is the natural way for them. Here Viveca is pulling out a wooden inspection board at the bottom of the hive to see how the bees are doing. Are there any dead bees? Or Varroa mites?

regarded by his devotees as something close to holy writ and thus beyond question.”

A few of Steiner's observations do coincide with reality, Chandler admits, and should therefore be heeded. This applies first and foremost to his warning against forcing bees into over-production.

Nor does he deny that there is a spiritual aspect to beekeeping. He himself has experienced periods of inner peace and "being present in the now" when he is with his bees that are much like when he meditates. "For the opportunity to experience that sense of timelessness in the company of a wild creature so many millennia our senior is a privilege beekeepers should celebrate and cherish."

EMMA AND JÖRN, TOP-BAR HIVE BEEKEEPERS
Lea, six, looks on as her parents Emma Bengtsson and Jörn Cordes inspect one of their hives. Although you run the risk of being called an incurable romantic as a top-bar hive beekeeper, that does not bother Emma and Jörn in the slightest. Their hives are ideal if you are keeping bees mainly for the sake of pollination.

Viveca Nilsenius is one of the few beekeepers in Sweden – or bee-carer as she calls herself – to have Warré hives. She used to keep bees commercially and on a large scale in expanded polystyrene hives until she read one of David Heaf's books on Warré beekeeping. This led her to completely change course.

The first time we met she was living on the Swedish island of Gotland and had fifty-five hives she had built herself. She had previously been a member of the local beekeepers' association, but she resigned after a meeting at which she was subjected to a great deal of pressure because of the methods she chose to employ. The worst she had done was to allow the bees to over-winter on their own honey and only harvested what was left in the spring.

"No-one came to my defence, though afterwards some members did tell me that what I was doing seemed interesting."

It wasn't just the way she harvested the honey that the others found so provocative, but also the fact that she allowed the bees and thus her entire apiary to propagate naturally by swarming. This meant that she did not exchange old queens for new ones fertilised by appropriate courtiers with the right genes, but allowed the virgin queens to fly off and mate freely with the drones in the neighbourhood.

"I believe in hybridising bees," she said. "It leads to genetic diversity and healthier bees."

It is not only in Sweden that Warré beekeepers may encounter a lack of sympathy. British Warréors have reported being mocked and ridiculed, though they also say that is changing. Conventional beekeepers are increasingly discovering that not everything alternative beekeepers say and do is completely absurd.

Viveca is currently living on the Bjäre peninusla in southern Sweden, where she is developing a new Warré apiary and running beekeeping courses. She, too, is seeing a growth in interest in alternative beekeeping methods, even if the situation is changing more slowly than in Great Britain.

Emma Svensson and Jörn Cordes are subsistence farmers in Bungemåla in south-western Småland. They decided to keep bees in order to improve the pollination of their crops and wild berries. They chose top-bar hives because they had read a book by Philip Chandler. The hives were at the perfect working height, and meant they could avoid lugging around boxes, frames and various bits of equipment.

After they had completed an intensive course with Patrick Sellman, who has made the top-bar hive better known in Sweden, they bought three kits. As soon as the hives were finished the bees were allowed to move in. Buckfast bees.

Twelve months have passed by the time I visit them. Everything has gone smoothly and it is hard to imagine nicer bees than theirs. You can even prod them without them showing the least concern. The family's protective overalls

PETER AND ÅSA, ORGANIC FARMERS
Most of Peter and Åsa Vingesköld's hives are frame hives in which the bees can construct their combs freely without wax foundations, but they also have some top-bar hives. Åsa is holding a top-bar comb, Peter a frame that is in the process of being filled.

seem completely superfluous, but I recognise myself from the time I started beekeeping. You get all the gear.

They don't have a smoker, though.

"The smoke makes the bees jittery," they say, "but if they start flying around and being difficult you can spray them with a little water."

Nor do they have one of the brushes other beekeepers use to remove the bees from the combs. Instead they use a goose feather the bees do not get stuck in. They make regular checks for any sign of Varroa mites, but fortunately they have only found them in such tiny quantities that they do not have to treat the hive. "If you protect the bees as a matter of routine, no natural selection can ever take place," they say. "And there is also the risk that the colonies are weakened because the Varroa treatments are so stressful."

To go with our coffee, Jörn serves a fragrant comb of fresh liquid honey straight from the hive. You chew on a bit of it and then spit out the wax that is left in your mouth. Delicious!

Peter Vingesköld, the co-author with Anette Dieng of the *Handbok i naturlig biodling* (The Natural Beekeeping Manual), the first Swedish book on the subject, has a great deal of experience in different methods of beekeeping. For twenty years he kept bees commercially and on a large scale. Then he made the move to organic beekeeping and received his certification according to the rules of K.R.A.V. (the Swedish organic certification organisation), which lay down that nectar and pollen sources within a three-mile radius have to be either K.R.A.V.-certified, E.U.-organic or natural in origin. But there was another change of direction after he attended a course on biodynamic farming followed by a move from Gnesta to the island of Gotland. He is the only beekeeper in Sweden who is Demeter-certified (see p. 132).

He tells me about the reaction to his book: it was completely panned in *Bitidningen* (The Bee Magazine), mostly because of his opposition to sugar feeding. "Though when I give talks to beekeeping associations I get a lot of positive responses from women and younger men."

Older men, on the other hand, find it difficult to accept his views. They have been feeding their bees with sugar for decades without doing them any harm. On the contrary, the bees have thrived on it, they say. The ability of bees to convert nectar into honey also means that they can break down sugar into fructose and dextrose, the substances that make up honey. A honey diet, on the other hand, can lead to the bees getting dysentery, particularly if the nectar source is heather or honeydew, which contains a lot of waste products.

To this Peter counters that honey has developed as bees and plants have evolved over forty million years, and ought therefore to be the best food for bees. He tells me that there is research to show that sugar deactivates more than two hundred of the genes in bees that affect their immune defences.

Right: An extraordinary collaborative effort is involved when a new comb is started in a top-bar hive. In this image a newly begun queen cell projects to the left, and pollen can be seen in the dark cells that previously contained the brood.

Left: in an almost completed comb the capped honey cells can be seen at the top.

Two beekeepers are capturing a swarm in a tree in order to place it in a hive, while two others open a hive that hangs from a tree. The text praises the virginal bees, their wax and honey. Illumination from the Barberini manuscript of the middle of the eleventh century.

POSTSCRIPT

MANY BOOKS come to a natural end, but for a while it seemed this book would never be finished. There was so much more to write about beyond what I had covered. I was keen to go to Poland to meet the tree beekeepers there. I wanted to go to Lithuania, where old traditions to do with bees and beekeeping have survived longer than in other European countries, and to Slovenia where beekeeping is called the "poetry of agriculture". I wanted to interview beekeepers who refuse to treat against the Varroa mite so their bees have the opportunity to develop a natural resistance. There was no end to the material that I felt it was important to include. But sometimes things happen that lead to major decisions making themselves.

I had the idea for this book several years ago, when John and Inga Larsson's granddaughter Anneli, who was visiting, told me about all the exciting things she had experienced while staying with her grandparents. And I realised that it was definitely time to wrap up the book when Helena, another grandchild, paid me a visit much later on with her father Bo. In this way the Larsson family came to frame the bee book project, and that felt entirely right.

After all, if I had not met John and Inga I would not have become a beekeeper, and in that case I could never have

Épisode du bal costumé donné par S. M. l'Impératrice (ballet des abeilles, l'arrivée des ruches).

The ballet of the bees, one of the highlights of a costume ball arranged by the Empress Eugénie in 1863. The hives are brought on and the queen bees peer out. From an illustration in Le Monde.

written this book. And if I had not written it I would never have realised how intertwined the lives of human beings and the honeybee have been through the ages, or how technical advances in the last one hundred and fifty years have fundamentally changed beekeeping. Nor would I have understood how powerful the economic interests are that threaten the survival of bees and other pollinators, and what a dreadful omen it was when the chemical giant Bayer bought Monsanto, another toxic giant, in 2018.

I would not have had the pleasure, either, of discovering the marvellous writers on bees of the past. And I wouldn't have got to know all the exciting bee people I have met over the years, including Marie Hannerstig, who has introduced live bees into her teaching courses at Vårfruskolan in Lund, and Robert Halling in Blentarp. It was at his apiary that I would for the very first time see honeycombs that the bees had constructed for themselves, without prefabricated wax foundations, an experience that has demonstrably affected the way I view bees and beekeeping. And I would never have gone to Læsø.

A huge thank you to everyone who has inspired, helped and encouraged me over the years I have spent working on this book.

Bee museums around the world

Australia	Illawarra Beekeepers, 98 Eton Street, Sutherlands, New South Wales
	Beechworth Honey, Historical Australian Archive and Museum, Ford Street, Beechworth, Victoria
Austria	Österreichisches Bienenzuchtmuseum, Orth an der Donau
Belgium	Tilff Bee Museum, Esplanade de l'Abeille 11, Tilff
	Musée de l'Abeille, Esneux, Liège
	Bijenteeltmuseum, Kalmthout, Antwerpen
Canada	Musée de l'Abeille et les Ruchers Promiel, Chateau Richer, Quebec
Finland	The Finnish Honeybee Museum, Juva
France	Le Musée du Miel, A Moure, Gramont, Gascogne
	Musée Vivant de l'Apiculture, La Cassine, Château-Renard, Loiret
	Le Musée de l'Abeille Vivante et la Cité des Fourmis, Kercadoret, LeFaouët, Bretagne
	Apiland Nature, Rousset, Bouches-du Rhône
Germany	Bienenmuseum, Moorrege, Schleswig-Holstein
	Deutsches Bienenmuseum, Weimar, Thüringen
	Bienenmuseum, Duisburg, Nordrhein-Westfalen
	Lebendiges Bienenmuseum, Knüllwald, Hessen
	Bienenkundemuseum, Münstertal, Schwarzwald
Greece	ΜΟΥΣΕΙΟ ΜΕΛΙΣΣΑΣ (Bee Museum), Pastida, Rhodos

Israel Bee Farm near Moshim Gan-Haim, Kfar-Saba

Italy Mulino Museo dell'Ape, Croviana, Trentino
Museo del Miele di Lavarone, Trentino
Museo di Apicoltura, Guido Fregonese, Oderzo, Veneto
Casa Museo dell' Apicoltura Tradizionale, Sortino, Sicilia

Lithuania Senovinės bitininkystės muziejus, Stripeikiai, Ignalina

Mexico Primer Museo de las Abejas en México, Colonia Condesa, Delegación Cuauhtémoc

Netherlands Bijenmuseum "De Bijenstal", Opperdoes

Norway Norwegian Beekeeping Museum, Billingstad

Poland Skansen Pszczelarski, Swarzedz, near Poznan

Slovenia Cebelarski muzej, Radovljica, near Lake Bled

Spain Casa-museo de Apicultura Ezkurdi, Eltso, Navarra
Museo de la Miel, Málaga
Museo de la Miel y las Abejas Rancho Cortesano, Jerez de la Frontera, Andalucia

U.K. Bee & Heritage Centre, Samlesbury Hall, Samlesbury, Preston, Lancashire
The Hive, Kew Gardens, London (wonderful hive sculpture in aluminium)
National Beekeeping Centre Wales, Bodnant Welsh Food, Furnace Farm, Tal-y-Cafn, Conwy

Ukraine Museum of Modern Beekeeping, Kiev

U.S.A. Honeybee Discovery Center, Orland, California
Museum of Fine Arts, St Petersburg, California ("Honey Bees at the M.F.A.")
U.S. National Pollinating Insects Collection, Logan, Utah
Ohio State University Bee Lab, Wooster, Ohio
American Museum of Natural History, New York, NY

Shops specialising in honey and honey products

Australia	Melita Honey Farm, Chudleigh, Mole Creek, Tasmania
	Beechworth Honey, Melbourne, Victoria
France	La Maison du Miel, Paris
	Le Comptoir du Miel, Paris. Branches in Lyon and Morzine, Haute Savoie
	Rucher de la Bouverie, Roquebrune-sur-Argens, Provence
Italy	La Casa del Miele di Simona Fregoni, Milano
New Zealand	Happy Valley, Rosehill, Auckland
	The Hive, Queenstown
	Honey Products New Zealand (online only – hpnz.nz)
Spain	El Colmenero, Madrid
	La Casa de Miel, Santa Cruz de Tenerife
U.K.	Bax Bees, Plumley, Cheshire
	Bee Baltic (online only – beebaltic.com)
	Bees for Development, Monmouth
	Bee Urban, The Hive, Kennington Park, London
	Bermondsey Street Bees, Bermondsey, London
	Chain Bridge Honey Farm, Horncliffe, Berwick-upon-Tweed

Local Honey Man, Walthamstow, London
Quince Honey Farm, Aller Cross, South Molton, Devon
Snowdon Honey Farm and Winery, Cadnant, Llanberis, Gwynedd
The Hive Honey Shop (online only – thehivehoneyshop.co.uk)

U.S.A.

Savannah Bee Company, Savannah, Georgia
Capital Bee Company, Savannah, Georgia
Carmel Honey Company, Carmel, California
Epic Honey Co. (online only – epichoney.com)
Sunny Honey Company, Seattle, Washington
The Beekeeper's Daughter, Plains, Pennsylvania
It's All About Bees!, Omaha, Nebraska

Bibliography

Andersson, Lars, *Bikungskupan* (Sodertalje: Forfattarforlaget, 1982)
Bergius, Bengt, *Tal om Läckerheter, både i sig själva sådana och för sådana ansedda genom Folkslags bruk och inbillning* (Stockholm: Natur och Kultur, 1960)
Birchall, Elizabeth, *In Praise of Bees, a Cabinet of Curiosities* (Shrewsbury: Quiller Publishing, 2014)
Bonsels, Waldemar, *Die Biene Maja und ihre Abenteuer* (Würzburg: Arena Verlag, 2005)
Butler, Charles, *The Feminine Monarchy, or the Historie of Bees* (London: 1623; facsimile editions: Northern Bee Books, 2010)
Chandler, Philip, *The Barefoot Beekeeper* (www. biobees.com, 2010)
Chandler, Philip, *Learning from Bees: A Philosophy of Natural Beekeeping* (Stockbridge, MA: Micro Publishing Media, 2012)
Clementz, L. J. Svensson, *En bok om bien och naturenligt rationell biodling* (Vallkarra: Eget förlag, 1902)
Crane, Eva, *The Archaeology of Beekeeping* (London: Duckworth, 1983)
Crane, Eva, *Honung* (Stockholm: Natur och Kultur, 1985)
Digges, Rev. J. G., *The Irish Bee Guide* (Leitrim: Irish Bee Journal Office, 1904)
Dutli, Ralph, *Das Lied vom Honig – Eine Kulturgeschichte der Biene* (Göttingen: Wallstein Verlag, 2012)
Fischerström, Johan, *Nya swenska economiske dictionnairen eller försök til et allmänt och fullständigt lexicon, i swenska hushållningen och naturläran* (Stockholm, 1780)
Fleischer, Esaias, *Udførlig afhandling om bier, og en for Dannemark och Norge Nyttig Bi-Avls Anlaeg* (Copenhagen: Lauritz Christian Simmelkiaer, 1777)
Gerner, P. Joh., *Handbok i rationell biskötsel* (Lund: Håkan Ohlssons boktryckeri, 1881)
Hallgren, Bengt, *Farfars honung* (Karlstad: NWT:s förlag, 1969)

Hansson, Åke, *Bin och biodling* (Stockholm: LT:s förlag, 1980)
Hansson, Åke, *Biodlingens grunder* (Stockholm: LT:s förlag, 1975)
Hansson, Åke (ed.), *Svensk Biodling* (Uppsala: Orbis, 1952)
Herwigk, Hans, *En nyttig bog om bier* (Copenhagen: 1649)
Hughes, Anne, *The Diary of a Farmer's Wife 1796–1797* (London: Penguin, 1981)
Imhoof, Markus & Lieckfeld, Claus-Peter, *More than Honey: The Survival of Bees and the Future of our World* (Berkeley/ Vancouver: Greystone Books, 2014)
Jones, Richard & Sweeney-Lynch, Sharon (ed.), *Collins Beekeeper's Bible* (London: HarperCollins, 2010)
Koch, Nils, *Swenska Bi-Skiötslen, upalstrad och efter mångfaldige rön samt enskildte kostsamma försök till fäderneslandets otroliga förmon* (Stockholm: Lorentz Ludwig Grefing, 1753)
Lagerlöf, Selma, *Gösta Berling's Saga* (London: Jonathan Cape, 1891, translated by Lillie Tudeer)
Laurel, Lars, *Den allmänna bi-skötslen äfter förfarenhet och försök I årdning ställd* (Lund: C. G. Berling, 1771)
Linnaeus, Samuel, *Kort, men tillförlitelig Bij-Skjötsel, på egen förfarenhet och anställte försök, efter bijens egentliga natur och egenskaper, grundad och inrättad, samt till allmänhetens tjenst och nytta, på mångas åstundan och anmodan* (Vaxjo: 1768/ Stockholm: Rediviva facsimile edition, 1975)
Ljungstrom, J. Alb., *Handbok i biskötsel i såväl halm-som ramkupor* (Stockholm: Albert Bonniers förlag, 1913)
Lundblom, Artur, *Honungsbiet i saga och sanning* (Stockholm: Natur och Kultur, 1959)
Lundgren, Alexander & Notini, Gosta, *Boken om bina* (Stockholm: Albert Bonniers förlag, 1943)
Maeterlinck, Maurice, *Bikupan* (Stockholm: Albert Bonniers förlag, 1902)
Marchese, Marina & Flottum, Kim, *The Honey Connoisseur* (New York: Black Dog & Leventhal Publishers, 2013)
Mattson, Carl Otto & Lang, Johann, *Bin till nytta och nöje*

(ArtCopy Sweden, 2009)
Milton, John, *The Practical Beekeeper Or, Concise and Plain Instructions for the Management of Bees and Hives* (London: 1851)
Natt och Dag, Gustaf, *Hufvud-Grunderne i Biskötslen för Enfaldige Landtmän* (Stockholm: 1784)
Petterson, Joachim, *Bisyssla* (Stockholm: Bonnier Fakta, 2015)
Preston, Claire, *Bee* (London: Reaktion Books Ltd, 2006)
Ramirez, Juan Antonio, *The Beehive Metaphor From Gaudi to Le Corbusier* (London: Reaktion Books Ltd, 2000)
Ransome, Hilda M., *The Sacred Bee* (New York: Dover Publications, 2004)
Rothman, Theodor Wolther, *Handledning wid bi-skötseln, ärnad til den okunnige bi-skötarens nytta* (Stockholm: 1800)
Scharp, Dag W., *Stora Biboken* (Malmo: Bengt Forsbergs förlag, 1966)
Seeley, Thomas D., *Honeybee Democracy* (Princeton & Oxford: Princeton University Press, 2010)
Steiner, Rudolf, *Bees: Nine Lectures on the Nature of Bees* (Herndon, VA: SteinerBooks, Inc., translated by Thomas Braatz, 1988)
Tavoillot, Pierre-Henri & Francois, *L'abeille (et le) philosophe* (Paris: Odile Jacob, 2015)
Triewald, Marten, *Nödig tractat om bij* (Stockholm: Andr. Bjorkman, 1728)
Trotzelius, Clas B., *Afhandling om skånska biskötslen* (Lund: C. G. Berling, 1759)
Vingeskold, Peter & Dieng, Anette, *Handbok i naturlig biodling* (Stockholm: Natur och Kultur, 2016)
Virgil, *The Georgics Book IV* (translated by Thomas Nevile, 1767)
Westberg, Sigurd (ed.), *Boken om bina* (Stockholm: Albert Bonniers förlag, 1943)
Wilson, Bee, *The Hive – The Story of the Honeybee and Us* (London: John Murray Press, 2004)

Index

Illustration Sources

Unterrichtswerk by Otto Schmeil
117 Heather Bees V © Laney Birkhead
118 © Alessandro Christiano/ iStockphoto
120 © Lotte Möller
122 Project Runeberg at Linköping University
124 Line drawing © E. H. Shepard, Reproduced with permission from Curtis Brown Group Ltd London, on behalf of The Shepard Trust
126 From *The ABC and XYZ of Bee Culture* by A. I. Root
126 Bohuslän Museum
129 Photographs by Wolfgang G. Vögele
130 NZ Museums
133 bpk/Stiftung Museum Schloss Moyland/Ute Klophaus/ © DACS 2020
134 Relief from the Church of St Peter am Wimberg
136 © Lotte Möller
137 Wikimedia Commons
140 Wikipedia
146 © 2018. Photograph Josse/ Scala, Florence
149 © Tom Gruat
152 Liebieghaus, Frankfurt am Main
154 (top) © Lotte Möller/(bottom) Photograph Uppvidinge Beekeepers Assocation
156 Photograph Anders Flood/ Västergötland Museum
161 From *The Honey Connoisseur* by Marina Marchese, Black Dog & Leventhal Publishers Inc
162 © Musée des HCL, Aurélie Troccon et Manon Mauguin
163 © Jens Ostman/KB
168 BnF Gallica
170 Upplands Museum (Paul Sandberg)
172, 175 © Lotte Möller
176 © Patricia Phillips/Alamy Stock Photo
178 Wikimedia Commons
183 http://brunbi.dk/fredning
184, 187 © Lotte Möller
190 Cueva de la Arana, Valencia, España
192 *Encyklopedia staropolska ilustrowana*, tr.1900, vol. 1, s. 122. (Z. Gloger)
193 Langstroth's *The Hive and the Honey Bee*. Courtesy of the Walter Havighurst Special Collections
195 Bibliothèque nationale de France
196 Örebro Läns Museum (Samuel Lindskog)
196 A Beekeeper, Gustav Johansson, St Anras (Fossemyr), Bohuslän Museum
198 The Wasatch Beekeeping Association
198 Sveriges Biodlares Riksforbund
200 Natural Beekeeping Trust
202, 203, 205, 207 © Lotte Möller
208 Biblioteca Vaticana, Cod. Barb. Lat. 592.
210 Bibliothèque nationale de France

Every effort has been made to trace the copyright holders of images not listed here. We would be glad to include any outstanding acknowledgements in future editions.

About the Author and Translator

LOTTE MÖLLER has worked for many years as a freelance writer and arts journalist. She is also the author of books on gardening and natural history, including *Trädgårdens natur* (The Nature of a Garden), *Tankar om trädgården* (Thoughts on a Garden) and *Citron*, a book about the lemon, which was shortlisted for the August Prize for Best Book of Non-fiction. *Bees and Their Keepers* was shortlisted for the August Prize in 2019. Lotte Möller lives in Lund, Sweden.

FRANK PERRY'S translations have won the Swedish Academy Prize for the introduction of Swedish literature abroad, the prize of the Writer's Guild of Sweden for drama translation and in 2017 his translation of Lina Wolff's *Bret Easton Ellis and the Other Dogs* was awarded the prestigious Oxford-Weidenfeld Translation Prize. He was awarded the triennial Bernard Shaw Prize in 2019 for best literary translation from Swedish.